INTRODUCTION TO BIOTECHNOLOGY

INTRODUCTION TO BIOTECHNOLOGY

K. S. Bilgrami
D.Phil, D.Sc, FNA
Department of Botany
Bhagalpur University, Bhagalpur

A. K. Pandey
Ph.D.
Department of Botany
Bhagalpur University, Bhagalpur

CBS
CBS Publishers & Distributors Pvt. Ltd.
New Delhi • Bengaluru • Chennai • Kochi • Kolkata • Mumbai
Hyderabad • Uttarakhand • Nagpur • Patna • Pune • Jharkhand

ISBN: 81-239-0133-X

First Edition: 1992
Reprint: 1994, 1998, 2001, 2003, 2004, 2005, 2007, 2008, 2015, 2019

Published by **Satish Kumar Jain** and produced by **Varun Jain** for
CBS Publishers & Distributors Pvt. Ltd.,
4819/XI Prahlad Street, 24 Ansari Road, Daryaganj, New Delhi - 110002
delhi@cbspd.com, cbspubs@airtelmail.in • www.cbspd.com
Ph.: 23289259, 23266861, 23266867 • Fax: 011-23243014

Corporate Office: 204 FIE, Industrial Area, Patparganj, Delhi - 110 092
Ph: 49344934 • Fax: 011-49344935
E-mail: publishing@cbspd.com • publicity@cbspd.com

Branches:
- ***Bengaluru:*** 2975, 17th Cross, K.R. Road, Bansankari 2nd Stage, Bengaluru - 70 • Ph: +91-80-26771678/79 • Fax: +91-80-26771680 E-mail: cbsbng@gmail.com, bangalore@cbspd.com
- ***Chennai:*** No. 7, Subbaraya Street, Shenoy Nagar, Chennai - 600030 Ph: +91-44-26681266, 26680620 • Fax: +91-44-42032115 E-mail: chennai@cbspd.com
- ***Kochi:*** Ashana House, 39/1904, A.M. Thomas Road, Valanjambalam, Ernakulum, Kochi • Ph: +91-484-4059061-65 Fax: +91-484-4059065 • E-mail: cochin@cbspd.com
- ***Kolkata:*** 6-B, Ground Floor, Rameshwar Shaw Road, Kolkata - 700014 Ph: +91-33-22891126/7/8 • E-mail: kolkata@cbspd.com
- ***Mumbai:*** 83-C, Dr. E. Moses Road, Worli, Mumbai - 400018 Ph: +91-9833017933, 022-24902340/41 • E-mail: mumbai@cbspd.com

Representatives:

- Hyderabad: 0-9885175004
- Patna: 0-9334159340
- Jharkhand: 0-9811541605
- Nagpur: 0-9021734563
- Pune: 0-9623451994
- Uttarakhand: 0-9716462459

Printed at:
J.S. Offset Printers, Delhi (India)

PREFACE

Biotechnology is a new and fast growing field of science. Apart from scientific applications toward deeper understanding of the nature and mode of regulation of genes, several broad area of application of recombinant DNA technology have been recognized in industry, agriculture and medicine. In view of the importance of Biotechnology, the University Grants Commission has emphasised on greater input of this discipline in the curricula of Plant Sciences at undergraduate level. The present book is based on the syllabus formulated by the Curriculum Development Committee of the UGC. As there is no text book available covering all the topics prescribed in the syllabus, the present book embodies a concise account of the fundamental aspects of plant cell and tissue culture, recombinant DNA technology and applications of biotechnology in medicine, human health, agriculture and industry. We hope that this book would provide basic information to the students on varied aspects of biotechnology.

We are thankful to Dr. H.C. Chaturvedi, Head, Tissue Culture Laboratory, National Botanical Research Institute, Lucknow and Prof. V.S. Jaiswal, Department of Botany, Banaras Hindu University, Varanasi for providing some of the photographs. Thanks are also due to Shri Santosh Thakur for preparation of drawings.

We would appreciate receiving suggestions for the improvement of the book.

K.S. Bilgrami

A.K. Pandey

CONTENTS

1

INTRODUCTION

HISTORICAL PERSPECTIVES

The biotechnological processes are actually in use since prehistoric times. The primitive people were well acquainted with alcoholic fermentation and vinegar formation from sugary substrates. The Sumarians were experts in brewing different types of beers. They had inscribed beer on a famous clay tablet, the "Monument Bleu" which is kept in the Louvre. There is also a number of documents through which it is established that as early as in the third millenium B.C., there existed a number of commercial breweries in Babylonia and Egypt.

Ethanol was first obtained in 1150 by distillation of alcoholic beverages. Since then it is being manufactured throughout the world. The use of leaven (an inoculum of mixed cultures of microorganisms transformed from one starter to another) in the fermentation of bread dough is known since many thousand years. This practice is still being used in bread industry.

In 14th century, first vinegar manufacturing industry was established in France near Orleans. Artificial cultivation of mushrooms was initiated in 1650 in France. In 1680 Antony van Leeuwenhoek first observed yeast cells with his newly designed microscope. Erxleben (1818) discovered the fermentive properties of yeasts. In 1857, Louis Pasteur highlighted the lactic acid fermentation by microbes. Hansen (1879) discovered the *Acetobacter* and Eduard Buchner (1897) reported that enzymes extracted from yeast can convert sugar into alcohol.

It was towards the end of the 19th century that large scale sewage purification systems employing microbes were established in Germany and France. Delbruck, Heyduck and Henneberg (1914 - 1916) discovered the large scale use of yeasts in food and fodder industry. During the same period, three important industrial chemicals (acetone, butanol and glycerine) were obtained from bacteria. Penicillin was discovered by Alexander Flemming in 1928 and large scale manufacture of penicillin started in 1944. Schatz and Waksman (1944) as well as Duggar (1944) discovered Streptomycin and Chlorotetracycline, respectively. From 1950 several new antibiotics have been discovered and added to the list of "wonder drugs".

In 1953, Watson and Crick revealed the structure of DNA which actually laid the foundation of genetic engineering. In 1962 mining of uranium with the help of microbes began in Canada. First successful genetic engineering experiment was done in 1973. In the middle and late eighties, several new antibiotics, and vaccines were developed. Interferon is being widely employed for the treatment of some viral diseases and even certain viral induced cancers.

In the history of biotechnology several stages of development can be recognised (Table 1.1)

Table 1.1. Historical Development of Biotechnology

Year	Development
Before 6000 BC	Yeast employed to produce beer and wine
Before 3000 BC	Providing bread with leaven, fermentation of juices to alcoholic beverages
3rd century BC	Manufacture of beer in Sumer and Babylonia and Egypt
From 1150 AD	Manufacture of ethanol
14th century	Vinegar manufacturing industry near Orleans
After about 1650	Artificial cultivation of mushroom in France
Before 1670	Copper mined with the help of micro-organisms
1677	Antony van Leeuwenhoek first observed yeast cells
1748	Observations on the generation, composition and decomposition of animal and vegetable substances
1818	Discovery of the fermentation properties of yeasts by Erxleben
1857	Louis Pasteur described Lactic acid fermentation
1869	F. Miescher isolated nucleic acid
1878	Joseph Lister reported lactic fermentation and its bearing on pathology

1879	Discovery of *Acetobacter*
1881	Production of lactic acid from microorganisms
1890	Alcohol first used to fuel motors
1890	S. Kitasato and E. Von Behring observed antibodies
1897	Eduard Buchner observed alcoholic fermentation without yeast cells
1910	Establishment of large scale sewage purification systems
1915	German process for the manufacture of bakery yeast
from 1920	Production of citric acid in a surface process
1928	Discovery of penicillin by Alexander Flemming
1944	Large scale production of penicillin, discovery of streptomycin
1946	Genetic recombination in bacteriophage
1953	Double stranded nature of DNA revealed
1962	Mining of Uranium with the help of microbes
1971	Transfer of *nif* gene from *Rhizobium* to *Klebsiella*
1973	First successful genetic engineering experiment done
1975	Discovery of hybridoma technique for production of monoclonal antibodies
1978	Restriction enzymes and their application to molecular biology
1981	US approval for using monoclonal antibodies in diagnosis
1982	Genetically engineered insulin approved for use
1984	Animal interferons approved for use against cattle diseases
Mid eighties	Use of interferons for the treatment of viral diseases
	Genetically engineered human growth hormone approved for treatment of dwarfism
	Wide use of monoclonal antibodies in diagnosis
	Introduction of genetically engineered hepatitis vaccine
	Production of new antibodies
Late eighties	Use of interferon for treatment of hairy cell leukaemia
	Raw materials obtained from microbes for plastic industry
	Large scale production of industrial chemicals using microbes

Nineties	Extraction of oil from the ground with the help of genetically engineered microbes
	Extraction of metals from factory waste
	Wide use of recombinant DNA technology in medicine, human health, agriculture and industry

PLANT CELL AND TISSUE CULTURE

Haberlandt (1898) was the first person to culture isolated vegetative cells of higher plants in simple nutrient solutions. Although Haberlandt was able to maintain the cells in these nutrient media, the cell division was not recorded until much later (Haberlandt, 1902).

Embryos were first plant tissue to be successfully cultured *in vitro* on artificial media. The first result in successful *in vitro* culture was by Hannig (1904) who grew under aseptic conditions relatively mature embryos of *Raphanus caudatus, R. landra, R. sativus* and *Cochlearia danica* (Brassicaceae) on Tollens medium with sugars, amino acids and plant extracts and obtained transplantable seedlings. Brown (1906), working with barley embryos, studied the relative efficiency of various organic nitrogenous compounds and reported glutamine to be the most effective in promoting shoot growth.

Andronescu (1919) observed that maize embryos when cultured without their endosperm, they gave rise to smaller plantlets than embryos cultured with their endosperm. Knudson (1922) succeeded in growing orchid embryos on nutrient medium in the absence of normal fungal symbionts. While working with embryos from a range of families and genera, Dietrich (1924) made two generalizations : (a) the embryos grown *in vitro* usually skipped a rest period and germinated, (b) the embryos taken into culture at less than one third of mature size were unable to produce viable plants capable of reaching maturity. Embryos from immature seeds tended to form malformed seedlings.

Laibach (1925, 1929) isolated embryos from non-viable seeds of the cross *Linum perenne* x *L. austriacum* and grew them to maturity on a nutrient medium. Schmucker (1929) grew individual mesophyll cells of *Bocconia* in a fluid prepared from a filtered extract of *Bocconia* leaves.

The first major breakthrough in the field of tissue culture was made in 1922 when Robbins (USA) and Kotte (a student of Haberlandt in Germany) achieved success in growing isolated root tips. Subsequently, Robbins and Maneval (1924) attained partial success in maintaining the culture for more than 20 weeks. White (1934) grew cultures of tomato root tips for a longer period using sucrose, inorganic salts and yeast extract.

The experiments on root culture raised many interesting questions. First of all, one wondered what could be the origin of the yeast extract activity used by Robbins (1922) and White (1934). The importance of thiamine in yeast extract was first demonstrated by Bonner (1937) and latter on confirmed by Robbins and Bartley (1937) and White (1937). This led to the fact that thiamine could be replaced by its two components, thiazole and pyrimidine. Thus it became possible to culture isolated roots on complementary synthetic media.

Further development in tissue culture technique was made in 1934 by Gautheret in France who cultured cambium cells of some tree species *viz.*, willow *(Salix capraea)* and poplar *(Populus nigra)* on Knop's solution. Nobecourt (1937) successfully cultured cambial tissue of carrot and tobacco for a longer period. This work was further extended by White (1939) who demonstrated the potentially unlimited growth of excised tomato root tips. Muir (1953) reported that by transferring callus tissue of *Tagetes* and *Nicotiana* to liquid medium and agitating the cultures on a shaking machine, it was possible to obtain single cells. Muir *et al.* (1954) further cultured single cells by nurse culture method and observed cell division.

Gautheret, White and Nobecourt are credited for laying the foundation for further work in the field of plant tissue culture. The tissue culture media, presently in use are modifications of those established by these three pioneer workers in 1939.

Morel and Martin (1952) for the first time obtained virus free *Dahlia* plants from diseased individuals by culturing healthy shoot tips. Morel (1960) also recovered virus free orchid plants.

The progress of tissue culture as a tool for the study of morphogenesis centres around the action of auxin on cell division and cell enlargement. Gautheret (1955) reported that callus tissues from different varieties of the same species may show different auxin requirement. Steward and Shantz (1956) demonstrated the capability of some of the externally applied auxins on cell division and cell enlargement. Patau *et al.* (1957) observed that cell division was initiated by the addition of IAA but for continuous growth, presence of a cytokinin was essential. Skoog and Miller (1957) reported that kinetin not only promotes cell division but also affects the growth of callus culture.

Skoog (1944) and Skoog and Tsui (1951) reported that in tobacco pith tissue cultures the addition of adenine and high levels of phosphate increased callus growth and bud formation even in the presence of IAA which otherwise acted as bud inhibitor.

The first report of formation of somatic embryos (commonly called embryoids) from tissues of carrot was published by Reinert and Steward (1958-59). Bergmann (1960) evolved the technique of cloning large number of single cells. He could get a population having 90% free cells by suspension culture method. Murashige (1961) demonstrated the technique of ***in vitro*** culture for propagation of various plant species.

Kohlenbach (1965) and Lang and Kolenbach (1975, 1978) demonstrated the ability of mechanically isolated, fully differentiated mesophyll cells of *Macleaya cordata* to yield an embryogenic callus. LaRue (1942) was the first to attempt the culture of ovaries of a few angiosperms and obtained root formation from the pedicels. Sachar and Guha (1962) cultured ovary of *Ranunculus scleratus* and induced polyembryony.

Kanta et al. (1962) cultured excised ovules and pollen grains together on the same medium. They observed that pollen grains germinated and fertilised the ovules. This technique was termed as "test tube fertilization".

The first pollen-derived haploid callus was obtained in anther cultures of *Tradescantia reflexa* (Yamada *et al.* 1963) but it failed to undergo organogenesis. Guha and Maheshwari (1964) reported development of embryos in anther cultures of *Datura innoxia* and subsequently confirmed their origin from pollen grains (Guha and Maheshwari, 1966). Bourgin and Nitsch (1967), using *Nicotiana* anthers observed that the pollen embryos developed directly into haploid plants.

Nitsch and Bui Dang Ha (1967) observed that the tissue proliferation will take place if they are subjected to the influence of auxin and kinetin in a specific order. Nitsch (1967) demonstrated that when GA is added with kinetin, it induces synergistic effect.

Another fascinating area of *in vitro* studies is the protoplast culture and somatic hybridization. Cocking (1960), for the first time, demonstrated the large scale isolation of protoplasts from root tips of tomato (*Lycopersicon esculentum*). Nagata and Takebe (1971) described for the first time, the regeneration of entire plantlets from mesophyll protoplasts of *Nicotiana tabacum*. Carlson *et al.* (1972) produced first somatic hybrid between *Nicotiana glauca* and *N. langsdorffii* by fusing their protoplasts. Since then several new hybrids have been obtained using the technique of protoplast fusion.

There is hardly any other field that has been revolutionized to such an extent as has been done by the advent of recombinant techniques. Weissmann and coworkers for the first time succeeded in coupling one of the IFN - genes with the *E. coli* genome to bring about interferon synthesis in bacteria (Nagata *et al.,* 1980). Since then recombinant DNA technology has travelled fast in providing the material and procedures for gaining insight into molecular processes and mechanisms for practical uses in diverse areas.

DEVELOPMENTS AND SCOPE

Biotechnology is a new and fast growing field of science. It includes branches like molecular biology, tissue culture, genetic engineering and plant pathology. Biotechnology is developing by close inter-weaving between these areas and can develop further only by means of an inter-disciplinary cooperation between them.

During the past decade, both basic and applied aspects of biotechnology have developed rapidly. The physico-chemical basis of biological systems and the control of their development and interactions are being unravelled in spectacular advances in molecular biology. Several new techniques have been developed which are providing valuable information in genetics, growth, development and reproduction of organisms.

The tools of plant cells and tissue culture are increasingly being applied to wide range of biotechnology ventures especially to the micropropagation and genetic improvement of crop plants. The techniques of protoplast culture and fusion are of special significance for improvement of crops such as potato, sugarcane, cassava and banana which are vegetatively propagated. Somatic cell fusions have been successfully employed for the resynthesis of rapeseed *(Brassica napus)* by fusing protoplasts of *Brassica oleracea* and *B. campestris* and for developing hybrids between *Medicago sativa* and *M. fulcata.*

Genetic engineering provides great promise for the improvement of crop plants. Genetically engineered crops with better nutritional status, resistance to insects, pests and herbicides, resistance to fungal, bacterial and viral diseases and resistance to frost damage have shown great potential. The techniques of biotechnology have already increased our capacity to enhance the crop production and in reducing the production cost. Substantial success has been achieved in genetic engineering of microbial systems. These systems can be endowed with specific genes to produce desired requirements of hormones, enzymes and vaccines. The vaccines produced by microbial organisms are sufficiently pure and specific for immunization against bacterial, fungal and viral diseases.

The techniques of recombinant DNA and gene cloning developed in the last decade have been used to demonstrate the ability to transfer genes between cells. The technology is sufficiently advanced so that potentially specific gene can be inserted into any plant cell. The insertion of any useful genes into otherwise productive crop species offers an opportunity to shorten the time requirement for the creation of new crop plants.

The commercial application of recombinant DNA technology has already begun and some of the important products which are available in the market include human interferons used for the treatment of leukaemia, growth hormones for the treatment of dwarfism, tissue plasminogen activator enzyme used against blood clot, human renin for the control of blood pressure and different types of vaccines against various diseases. Recombinant DNA technology is also being used in the production of monoclonal antibodies.

The development of techniques of DNA transfer to plant by using the Ti plasmid of *Agrobacterium* and other vectors is an important landmark in plant genetic engineering. Biological control of agricultural pests by microbes has also been found to be highly promising. The introduction of the *Bacillus thuringiensis* gene for insect resistance into tabacco and tomato and the genes for virus resistance into different species are the major achievements in the field of biotechnology. In future Biotechnological developments are likely to revolutionise the medical, agriculture and other plant-based industries.

TERMINOLOGY

Biotechnology

There is considerable diversity regarding the exact definition of biotechnology. Term biotechnology is composed of two words-bio (Greek *bios* - means life) and technology (Greek *technologia* - means systematic treatment). In other words, it is the science of applied biological processes. According to the Oxford Dictionary, technology is the "scientific study of the practical or industrial arts" or the "technology of a particular art or subject". Webster's Dictionary, however, defines technology as "the application of scientific knowledge to practical purposes in a particular field" and biotechnology as "the aspect of technology concerned with the application of biological and engineering data to problems relating to the mutual adjustment of man and the machine".

Spinks (1980) defined biotechnology as "the application of biological organisms, systems or processes to manufacturing and service industries".

Biotechnology may also be defined as "the science of the production processes based on the action of microorganisms and their active components, and of production processes involving the use of cells and tissues from higher organisms" (cf. Biotechnology : A Dutch Perspective, 1981).

European Federation of Biotechnology (1981) defined biotechnology as "the integrated use of biochemistry, microbiology and engineering sciences in order to achieve technological (industrial) application of the capabilities of microorganisms, cultured tissue cells and parts thereof".

In the report Biotechnology : A Development Plan for Canada (1981) biotechnology is defined as "the utilization of a biological process, be it microbial, plant or animal cells, or their constituents, to provide goods and services". On the other hand, Biotechnology for Australia (1981) regarded biotechnology as "the devising, optimising and scaling up of biochemical and cellular processes for the industrial production of useful compounds and related applications".

Bull *et al.* (1982) defined biotechnology as "the application of scientific and engineering principles to the processing of materials by biological agents to provide goods and services".

Genetic Engineering

Genetic engineering (often referred to as recombinant DNA technology) may be defined as "the experimental manipulation of DNA (or RNA) of different species producing recombinant DNA which includes some genes from both species".

2

TISSUE CULTURE MEDIA

Plants grown in nature get their nourishment from nutrients available in soil and also synthesised by the plant tissues themselves. Plants when grown *in vitro,* solely depend on the nutrients contained in the medium. Nutritional requirement of different species may vary, therefore,a wide range of culture media differing in their elemental compositions have been formulated for culture of plant cells/ tissue (Table 2.1).

Tissue cultures can be maintained either in liquid medium or semi-solid medium. In the former, plant material is immersed in the medium either partially or completely while in the latter plant material is placed on the surface of the medium.

Suspension cell cultures are initiated by transferring preferably friable callus to liquid medium containing organic and inorganic nutrients and growth hormones. Suspension cultures are regularly aerated either by bubbling sterile air or gentle agitation on an orbital shaker at between 90 and 150 rpm, with an orbital diameter of 1-2 cm. The culture of plant tissues in an agitated liquid medium facilitates gaseous exchange, removes any polarity of the tissue due to gravity and eliminates nutrient gradients within the medium and at the surface of the cells.

COMPOSITION

Cells, tissues and organs can only be cultured *in vitro* when all the required nutrients are available in the medium. A nutrient medium should contain inorganic nutrients, organic nutrients and growth hormones.

1. INORGANIC NUTRIENTS

A number of inorganic nutrients are required for normal growth of the plants. Besides, carbon, hydrogen and oxygen, other 12 elements essential for plant growth include nitrogen, calcium, potassium, phosphorus, sulphur, magnesium, iron, manganese, copper, boron, zinc and molybdenum. Out of these, former six elements are macronutrients as these are required in comparatively larger quantities (concentrations more than 0.05 mmol l^{-1}), while the latter six are classed as micronutrients as these are required in smaller quantities (less than 0.05 mmol l^{-1}). In White's medium iron is added in the form of Ferric sulphate ($Fe(SO_4)_3$) but in most of the other tissue culture media iron is supplied in the form of Fe. EDTA. Nitrogen is furnished in the medium in the forms of nitrate (oxidised) and ammonia (reduced).

Deficiencies of the macroelements N,P,S,K, Mg and Ca are more easily manifested in tissue culture when cells are cultured in liquid rather than on agar media since impurities present in the agar are considerable. Of all the mineral nutrients, the form of nitrogen is responsible for the most pronounced effects on growth and differentiation of cultured tissues.

2. ORGANIC NUTRIENTS

Cultured plants accomplish better growth when medium is supplemented with organic nutrients such as amino acids and vitamins. The most commonly used vitamin is thiamine (vitamin B_1). Other vitamins which improve the growth of cultured plants include nicotinic acid, calcium pantothenate and pyridoxine. Besides these, biotin, folic acid and aminobenzoic acid are added in various concentrations in some of the media.

As a carbon source, sucrose is the most preferred carbohydrate. Glucose, fructose, maltose, galactose, mannose and lactose are the other favourable sugars.

In order to promote growth of callus, some complex semisynthetic substances are added in the medium. These include yeast extract (YE), coconut milk (CM), casein hydrolysate (CH), tomato juice (TJ) and malt extract (ME).

3. GROWTH HORMONES

In addition to inorganic nutrients, vitamins and the carbon source, basal medium is supplemented with growth hormones such as auxins, cytokinins and gibberellins.

TABLE 2.1 Constituents of some common plant tissue culture media (mg litre^{-1}).

Constituents		WM	MS	NM	NT
Inorganic compounds					
Potassium chloride	KCl	65	-	-	-
Potassium nitrate	KNO_3	80	1900	950	950
Ammonium nitrate	NH_4NO_3	-	1650	720	825
Calcium chloride	$CaCl_2.2H_2O$	-	440	166	220
Calcium nitrate	$Ca(NO_3)_2$	300	-	-	-
Magnesium sulphate	$MgSO_4.7H_2O$	720	370	185	1233
Sodium sulphate	$Na_2SO_4.10H_2O$	200	-	-	-
Potassium dihydrogen phosphate	KH_2PO_4	-	170	68	680
Sodium dihydrogen phosphate	$NaH_2PO_4.7H_2O$	165	-	-	1233
Manganese sulphate	$MnSO_4.4H_2O$	3	22.3	25	22.3
Boric acid	H_3BO_3	1.5	6.2	10	6.2
Zinc sulphate	$ZnSO_4.7H_2O$	3	8.6	10	8.6
Potassium iodide	KI	0.75	0.83	-	0.83
Sodium molybdate	$Na_2MoO_4.2H_2O$	-	0.25	0.25	0.25
Copper sulphate	$CuSO_4.5H_2O$	0.01	0.025	0.025	0.025
Cobalt chloride	$CoCl_2.6H_2O$	-	0.025	-	-
Cobalt sulphate	$CoSO_4.7H_2O$	-	-	-	0.03
Disodium ethylene diamine tetra-acetate	$Na_2EDTA.2H_2O$	-	37.3	37.3	37.3
Ferrous sulphate	$FeSO_4.7H_2O$	-	27.8	27.8	27.8
Ferric sulphate	$Fe(SO_4)_3$	2.5	-	-	-

(Contd.)

Constituents	WM	MS	NM	NT
Organic compounds				
Inositol	-	100	100	100
Nicotinic acid	0.05	0.5	5	-
Pyridoxine HCl	0.0	10.5	0.5	-
Thiamine HCl	0.01	0.1	0.5	1
Glycine	3	2	2	-
Ca-D-pantothenate	1	-	-	-
Sucrose	20,000	20,000	20,000	20,000
Mannitol	-	-	-	13%w/v
Growth regulators				
IAA	-	1.30	-	-
NAA	-	-	-	3
Kinetin	-	0.4-10	-	-
6 benzyl amino purine	-	-	-	1

WM - White's medium (1954)

MS - Murashige and Skoog medium (1962)

NM - Nitsch' medium (Nitsch, 1969)

NTM - Nagata and Takebe's medium (Nagata and Takabe, 1972)

Auxins promote cell division and root differentiation. Most commonly used auxins are: IAA (indole acetic acid), IBA (indole-3-butyric acid), NAA (naphthalene acetic acid) and 2, 4-D (dichlorophenoxy acetic acid). Cytokinins are responsible for cell division and shoot differentiation. BAP (benzylamino purine), and 2-isopentanyl adenine are most widely used cytokinins. Among gibberellins, GA_3 is commonly used.

4. AGAR

Agar (a polysaccharide obtained from seaweeds) is used to solidify the medium. Agar is commonly used at a concentration of 0.8-1% (w/v). Use of higher concentration of agar makes the medium hard. This prevents the diffusion of nutrients into the tissues.

5. pH

The pH of the medium is adjusted between 5 to 5.8 by adding (O.1N) NaOH or HCl. Usually a pH higher than 6 results in a fairly hard medium whereas a pH below 5 does not allow satisfactory solidification of the medium.

MEDIA PREPARATION

The simplest method of preparing culture media is to use commercially available powdered media. These media contain all the requisite nutrients. However, agar, sugar and other supplements are added in the finally prepared media.

Another method of preparing media is to prepare concentrated stock solutions. For the preparation of stock solutions, distilled water and chemicals of high purity are used. Stock solutions varying in strength are prepared of different media components viz., inorganic salts, vitamins and other organic compounds and growth regulators. All the stock solutions are stored either in glass or plastic containers and stored in refrigerator. Stock solution of iron is stored in amber-coloured bottles. Substances which are not stable in frozen state must be freshly prepared each time the medium is required and added to the final mixture of stock solutions.

For the preparation of medium, the following sequencial steps are followed:

1. Appropriate quantities of agar agar and sucrose are dissolved in distilled water.
2. Required quantities of stock solutions, growth hormones and other supplements are added.
3. More distilled water is added to make the final volume of the medium.
4. pH of the medium is adjusted between 5 - 5.8 by adding 0.1N NaOH or 0.1N HCl.
5. The medium is poured in culture tubes, flasks or any other containers.
6. Culture vessels are plugged with non-absorbent cotton wool wrapped in cheese cloth. This will allow free gaseous exchange but inhibit microbial contamination.
7. Culture vessels are autoclaved at 120°C (1.06 kg/cm^2) for 15-20 minutes.
8. Culture medium is allowed to cool at room temperature and used or stored at 4°C.

Media Selection

As there is no single medium suitable for culturing all cell, tissue and organs, it is necessary that the most suitable medium be determined to accommodate specific requirements of the plant material in question. It is always better to begin with well known basal medium (Table 2.1). To select a suitable medium for an untested system, broad spectrum experiments may be setup as suggested by De Fossard *et al.* (1974). In this experiment all the components of the medium are divided into four major groups: (a) minerals, (b) auxins, (c) cytokinins and (d) organic nutrients. For each major group, three concentrations (low, medium, high) are selected. Various combinations of four groups and three concentrations result in most suitable media for a particular species.

STERILIZATION

The media used for plant tissue cultures also provide suitable environment for the growth of microorganisms like bacteria and fungi. These organisms grow faster than the cultured tissues and produce toxic metabolites which inhibit the growth of explant or even kill it. Therefore, it is absolutely essential to provide and maintain a completely aseptic environment inside the culture vessels.

Sterilization Room

The dimensions of sterilization room should be 6x6x4.5 sq.m. The room should be equipped with an autoclave, either of horizontal or vertical type. Usually former type of autoclaves are preferred. The room should have a good ventilation.

Sources of Contamination

There are a number of sources through which medium may get contaminated. These include medium itself, explant, culture vessels and instruments, environment of culture room and place of transfer.

(a) **Medium:** The microbes may be present in the medium right from the very beginning. To eliminate them, vessels containing the medium are plugged and autoclaved at 120°C for 15-20 minutes. Sterilization is done in steel autoclaves or in pressure cookers if former is not available.

Some growth substances and vitamins are thermo-labile. Thus they can not be autoclaved. Such compounds are sterilized by passing through membrane filters of 0.22 μm pore size. Small volumes can be sterilized using syringe holder but large volumes (above 500 ml) require vaccum filteration.

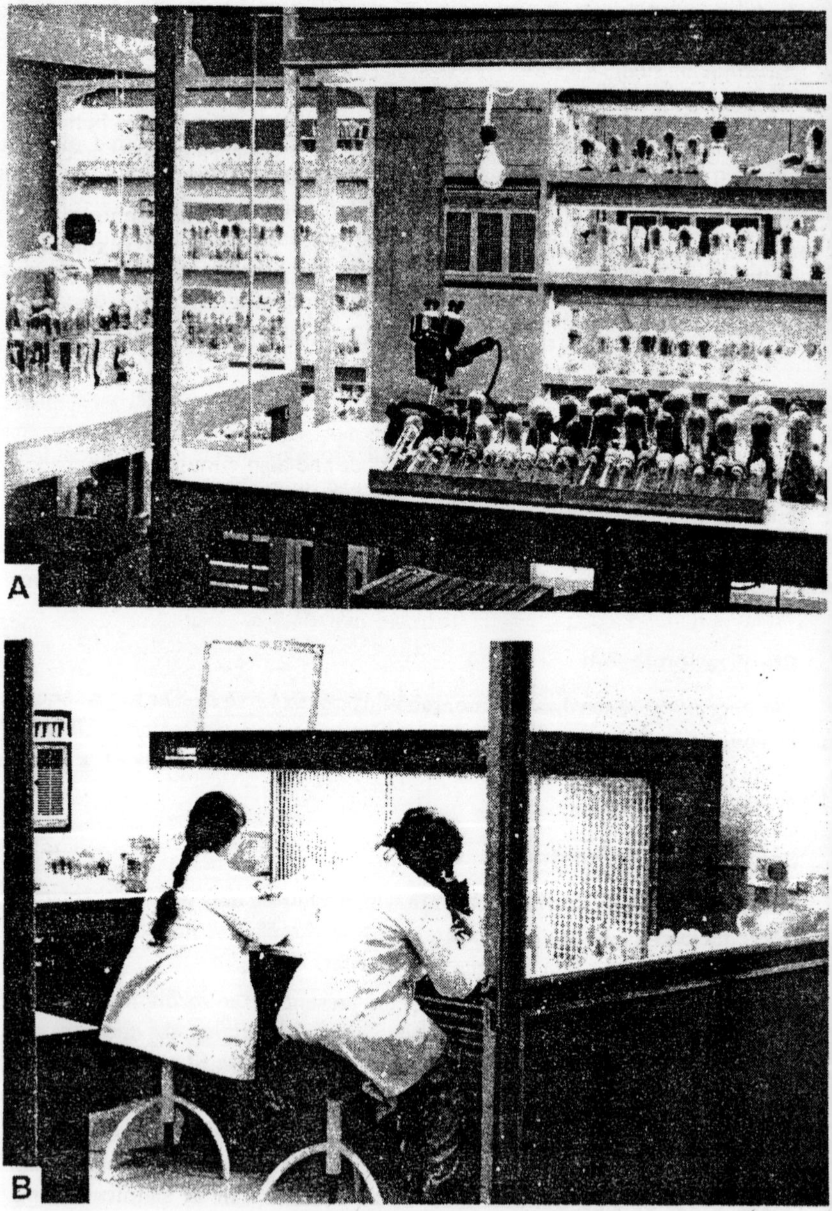

Fig.2. A. A view of one of the culture rooms of the Tissue Culture Laboratory, National Botanical Research Institute, Lucknow, **B.** Inoculation work being performed in a laminar flow cabinet.

(b) **Culture vessels and instruments**: Glasswares and metal instruments are sterilized by placing them in an oven at 160-180°C for 3 - 4 hours. The instruments such as scalpel, forceps, needles etc., dipped in 95% ethyl alcohol, are flammed before use.

(c) **Plant material:** The plant tissue which is to be cultured must be surface sterilized before placing it on the medium. This is essential to avoid microbial contamination. Usually calcium or sodium hypochlorite solutions are used. In cases where plant material is very hard such as seeds, treatment with disinfectant may be done directly. On the other hand, while culturing young embryos or ovules, surface sterilization of protecting tissues (e.g. ovules in case of embryos and ovary in case of ovules) are done to avoid damage to the explant.

(d) **Transfer area:** Aseptic manipulations are conducted in a closed transfer area fitted with gloves and UV lamps. Ultra violet light sterilizes the interior portion of the inoculation chamber and also eliminates atmospheric contamination. These days, in most of the tissue culture laboratories, laminar air flow cabinets are used. Air passes through ultrafilters which flows in the working area at a constant speed. These filters not only eliminate dust and other particles but also fungal and bacterial spores. Thus an aseptic environment is maintained at the time of tissue inoculation.

3

CELLULAR AND TISSUE DIFFERENTIATION

The process of the growth accompanied with morpho-physiological specialization of the cells produced by the meristems is called **differentiation**. In tissue culture, the term differentiation is used to describe the formation of different cell types and organs in the callus or cell culture.

After differentiation, those tissues which loose their meristematic activity are called permanent tissues. These highly mature and differentiated cells possess the ability to grow, divide and redifferentiate under appropriate experimental conditions. If tissues have gone irreversible differentiation, such tissues would not be able to divide further. Thus all the living cells in a plant body, irrespective of their ploidy level and the form of specialization, are capable of giving rise to whole plants. This inherent property of cells is called **cellular totipotency**.

In both plants and animals specialised cells perform different functions. This specialization is known as **cytodifferentiation**. Already differentiated cells may acquire meristematic activity forming callus tissue. This process of reversal of mature cells to meristematic activity is called **dedifferentiation**. The ability of dedifferentiated cells to form a whole plant or plant organ is known as **redifferentiation**.

Anatomical studies of the callus tissues have revealed that they are not entirely composed of dividing cells but, instead, show restricted meristematic regions dispersed in a ground tissue. The cells of a callus show high degree

Table 3.1. Degree to which different cell types may dedifferentiate (after Gautheret 1966)

Vegetative point	Cambium	Companion and secretory cells	Parenchyma	Thick-walled cells (collenchyma, lignified cells)	Fibres	Degenerating cells (vascular elements sieve tubes)

of vacuolation. The callus cultures always possess a degree of internal organization and differentiation. Gautheret (1966) observed that the degree of regression a cell can undergo would depend on the cytological and physiological state it has reached *in situ* (Table 3.1).

Cellular totipotency reveals the fact that all the necessary genes responsible for differentiation are present in a somatic cell but many of them do not express themselves in differentiated tissue or organs. Such inactive genes may become active when differentiated cells are cultured on appropriate medium.

Growth and differentiation, although proceed together but they are the independent processes. Growth may occur without differentiation e.g. callus tissue. Likewise, differentiation may be accomplished without growth e.g. early stages of many embryos. Sometimes, culture conditions may induce either of the processes. Abundance of water and available nitrogen tend to induce growth whereas abundance of carbohydrates and shortage of nitrogen and water promote differentiation. Growth usually ceases when differentiation process is complete.

Differentiation may be categorised into two groups: structural and physiological.

1. **Structural differentiation**: Structural differentiation is further distinguished into external and internal differentiation.

External differentiation: The most common example of external differentiation is root and shoot differentiation in higher plants. Differentiation of these two organs vary in structure, function and pattern of growth. Another familiar instance of differentiation is vegetative and reporductive phases of life cycle. Further differentiation in reproductive organs results in male and female organs.

Internal differentiation: This includes differentiation of various types of cells and tissues. The differentiation of one type of cell may induce changes in the character of adjacent cells. The position of the cell in the developing system is closely related to the type of differentiation it undergoes. Differentiated cells mostly occur in groups forming different type of tissues e.g. vascular tissues.

2. **Physiological differentiation**: The variations in the structure between root and shoot are the expressions of fundamental physiological differences. Roots and shoots vary in their ability to synthesise substances like vitamins, alkaloids etc. In cultures, roots may grow indefinitely from a small piece of root tip but such cultures must be supplemented with small quantity of certain vitamins especially thiamine. In nature, thiamine is

synthesised in shoots and is made available to the root of the plant. Torrey (1953) observed that the synthetic substances which inhibit root elongation have specific effect on acceleration or retardation of the differentiation of xylem and phloem.

VASCULAR TISSUE DIFFERENTIATION

Among vascular tissues, xylem is the most complex one as it comprises several types of cells. The most important cells are the tracheary elements which are principally concerned with the transport of water. The tracheary elements are distinguishable into tracheids and vessels. The main difference between tracheids and vessel members are that the former are not perforated while the end walls of the latter are perforated.

The callus cultures offer a valuable system for the study of cytodifferentiation *in vitro*. The parenchymatous cells of the callus differentiate into the cells having secondary wall thickenings as observed in xylem (Fig. 3.1.). For the study of cytodifferentiation, more attention has been paid to the tracheary elements of xylem because of the following advantages:

1. Tracheary elements can be easily stained, thus it is easier to locate them in macerated preparations.
2. Tracheary elements have different types of wall thickenings viz., annular, spiral, reticulate or pitted. Thus it is easy to identify differentiated cells from those of undifferentiated cells.
3. The development of tracheary elements can be induced in cell and tissue cultures by controlling the cultural conditions.
4. Cytodifferentiation of tracheary elements is an example of irreversible specialization.
5. It is easy to get marker proteins in tracheary elements during cytodifferentiation.

During cytodifferentiation of tracheary elements, a particular cell acquires capacity for differentiation. This is followed by the process of cell enlargement, secondary wall development, and lignification. After completion of secondary wall thickenings and their lignification, the protoplast and certain part of the cell wall enter the stage of lysis. In cultured tissues, cytodifferentiation is triggered by growth hormones and sucrose.

In vascular tissue differentiation graft union is not necessary. This has been demonstrated in cultured callus of *Syringa vulgaris* by Wetmore and Sorokin (1955). They made a v-shaped incision in the callus and grafted a bud of the same species. Immediate induction of mitosis resulted in the formation of short vertical column of cells which differentiated to form lignified and

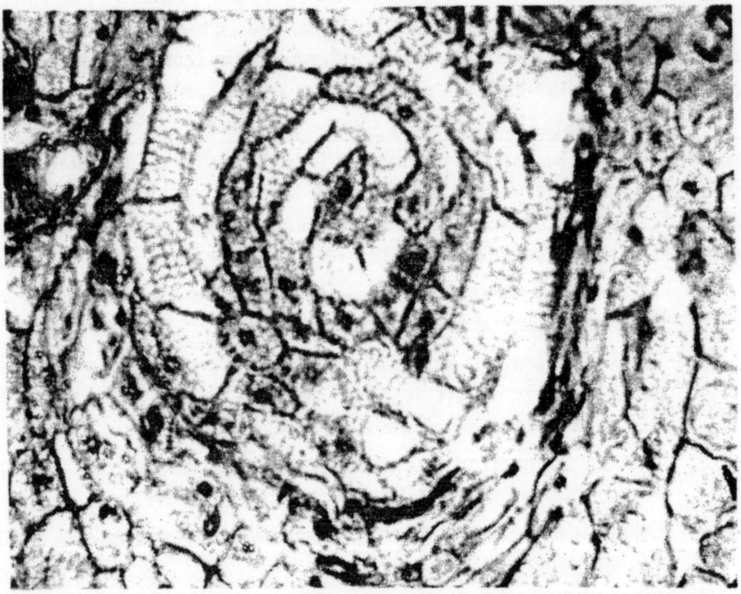

Fig.3.1. Cross-section of somatic callus of *Rauvolfia serpentina* showing differentiation of tracheary elements as concentric nests amongst the parenchymatous cells of callus.

pitted xylary elements (Fig. 3.2). Vascularization enhanced when more auxin was added to the medium. Further increase in the auxin concentration, altered the position of the formation of vascular elements away from the place of grafting. Tracheary element differentiation takes place even if the contact between the callus and the bud is broken by placing a cellophane paper at the site of grafting. This clearly indicates that stimulus for differentiation is of diffusible nature.

FACTORS AFFECTING CYTODIFFERENTIATION

1. Phytohormones

Auxin at low concentrations stimulates xylem differentiation. An increase in the concentration of auxin alter the position of the vascular strands. This has been demonstrated in grafting experiments of *Syringa vulgaris.* Cytokinin and gibberellin also stimulate tracheary element differentiation. When auxin and kinetin are used together, they have a synergestic effect as has been observed in tobacco and *Zinnia.* An inhibitory effect of kinetin has been observed in wound vessel formation in *Coleus* stem.

2. Sucrose

Sucrose plays an important role in vascular tissue differentiation. Its concentration directly affects the relative amounts of xylem and phloem formed in the callus tissues in the presence of low concentration of auxin. This has been demonstrated in the experiments done on *Syringa* and *Phaseolus.* In *Syringa* when agar containing 0.05 mgl^{-1} IAA and 1% sucrose was applied to the cavity at the top of the callus, only a few xylem elements could develop. Better xylem differentiation (with little or no phloem) was observed when sucrose level was increased to 2%. Both xylem and phloem differentiated at 2.5 - 3.5% sucrose concentration. With an increase in sucrose concentration (4%) phloem was formed with little or no xylem.

3. Nitrogen

Presence of ammonia and nitrate in the medium directly affects the differentiation of tracheary elements. Reduction in the total nitrogen (NH_4^+ and NO_3^-) content of the medium increases the percentage of tracheary elements in Jerusalem artichoke *(Helianthus tuberosus).* In *Zinnia* low concentration of NH_4^+ in the medium stimulates cytodifferentiation but higher concentrations reduce the formation of tracheary elements.

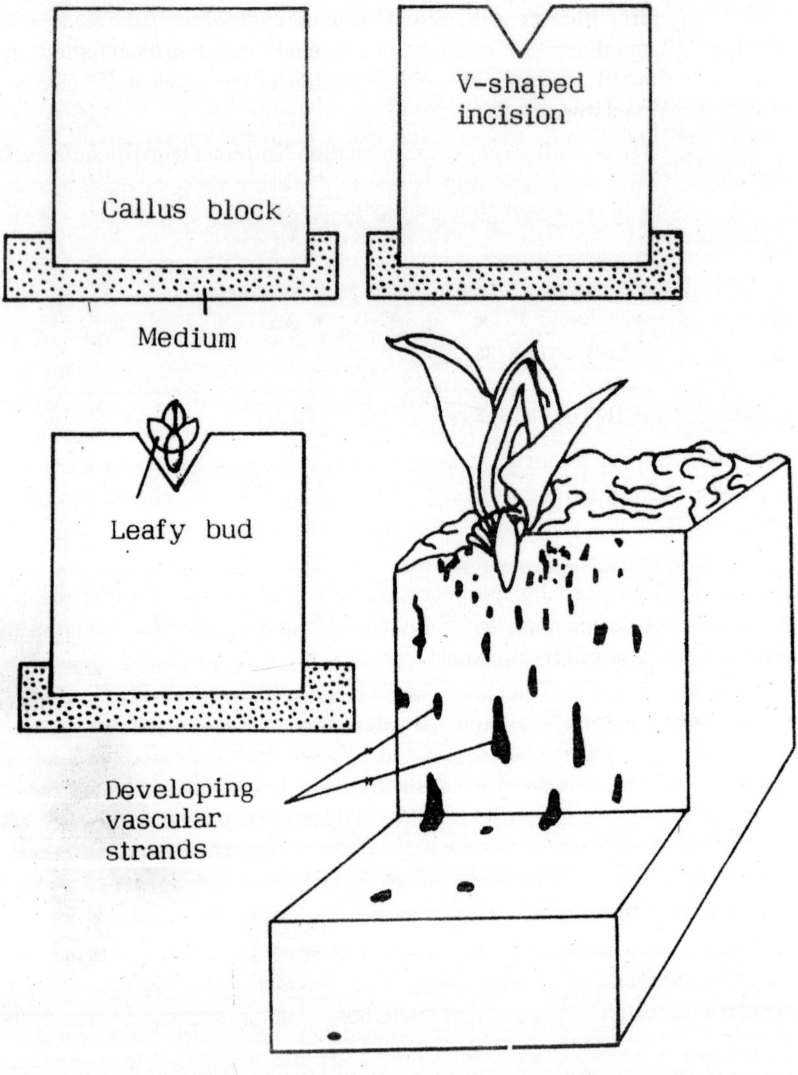

Fig.3.2. Introduction of vascular elements differentiation in callus tissue of *Syringa vulgaris*.

4. Physical Factors

The effect of light on vascular tissue differentiation varies between cultured tissues. In carrot phloem slices, light induces cytodifferentiation via the synthesis of cytokinin. Light stimulates wound vessel differentiation in *Coleus*. In *Sinapis* tracheary element formation in the hypocotyle is controlled by phytochrome.

Temperature also affects cytodifferentiation. In Jerusalem artichoke callus cultures grown at a temperature below 17°C did not show vascular strands but an increase in temperature within the range of 17-31°C promotes vascular differentiation.

pH of the medium greatly affects cytodifferentiation. In *Zinnia* the optimum pH was found to be 5.0. pH below and above 5, results in the decrease of tracheary elements differentiation.

CYTODIFFERENTIATION AND CELL CYCLE

Fosket (1968) for the first time suggested a close relationship between tracheary element differentiation and the cell cycle. He observed that in cultured *Coleus* stem segments inhibitors of DNA synthesis (mitomycin C, fluorodeoxyuridine) and inhibitor of mitosis (colchicine) prevented cytodifferentiation. In such cases cell division preceded the appearance of xylary elements. This led Fosket to conclude that cell division is essential for the induction of tracheary element differentiation.

The idea that cell division is not always a prerequisite for tracheary element differentiation was putforth by Torrey (1975). He observed that in cell cultures of *Centaurea cyanus* some of the single parenchymatous cells directly gave rise to tracheary element without cell division. The work of Torrey got support from the findings of Kohlenbach and Schmidt (1975) and Fukuda and Kommine (1980) who reported that mature mesophyll cells of *Zinnia elegans* can differentiate to tracheary elements without undergoing cell division.

It has been reported that DNA replication is not necessary for cytodifferentiation but some studies have suggested the involvement of minor DNA synthesis in differentiation. The regulation of gene expression has been studied using an inhibitor 5-Bromodeoxyuridine. Tracheary element differentiation from parenchyma cell is prevented by this inhibitor in pea and *Zinnia* but cell division is not affected. This further indicates that replication of whole genome and synthesis of specific DNA sequences are essential for cytodifferentiation. In cultured cells with the increase in endopolyploidy, differentiation ability of cells may decline.

4

CALLUS GROWTH PATTERNS

Callus may be defined as an unorganised mass of loosely arranged parenchymatous cells which develop from parent tissue due to proliferation of cells (Fig. 4.1).

CALLUS INITIATION

Callus initiation begins when segments (explants) of plant organs are cultured on solidified culture media. Callus cultures can be initiated from explants taken from any part of the plant such as root, stem, leaf, seed etc. Such cultures can be maintained for an indefinite period if they are transferred to fresh nutrient media at regular intervals. This is known as subculturing.

Explants from carrot root have been most widely used to demonstrate callus initiation. Gautheret (1959) observed that the callus formation by carrot root tissues is related to the region from which the explant has been removed. For example, explants of xylem and phloem adjascent to the cambium region show vigorous growth and produce enormous amount of callus.

A procedure for the initiation of callus culture from tobacco stem tissue is outlined in figure 4.2.

When explant is placed on the medium, cellular masses begin to appear around the cut edges of the explant. The initiation of callus formation is visible when the surface of the explant begins to glisten and the texture becomes rough. Repeated cell divisions cover the entire explant with unor-

Fig.4.1. A typical friable callus-habituated stem callus of *Citrus grandis.*

ganised callus in 2-3 weeks (Fig. 4.3). The cell division increases the size of the explant. If the callus is allowed to grow on the same medium for a longer period, depletion of nutrients and desiccation of agar will occur which will affect the further growth of the callus. Besides this, developing callus may also produce metabolites which may have toxic effect. This may be overcome by subculturing the callus. The callus is divided into 2-4 equal segments and transferred onto fresh culture medium of the same composition. After several such subcultures, callus gets established and its growth gets stablized.

Stem segment
↓
Sterilization
↓
Rinsing in distilled water
↓ Cut
Stem discs
↓
Plating on agar solidified medium
↓
Callus

Fig.4.2. Initiation of callus.

Callus cultures may be initiated from any meristematic tissue but all callus cultures do not produce plants. In grasses, however, callus cultures derived from immature embryos show highest rate of plant regeneration.

Callus cultures which comprise unorganised mass of cells may be subcultured to produce more calli or may be induced to produce roots, shoots or embryoids which further get differentiated to form new plants.

In cereals (wheat, rice, maize and sorghum) seedling segments may be induced to produce callus. These cultures initially remain unorganised but subsequently develop roots. When callus growing on agar medium was transferred to liquid medium of same composition, a large number of roots with lateral roots developed. But when this differentiating tissue was transferred to the agar medium, they reverted back to unorganised tissue and further developed as callus.

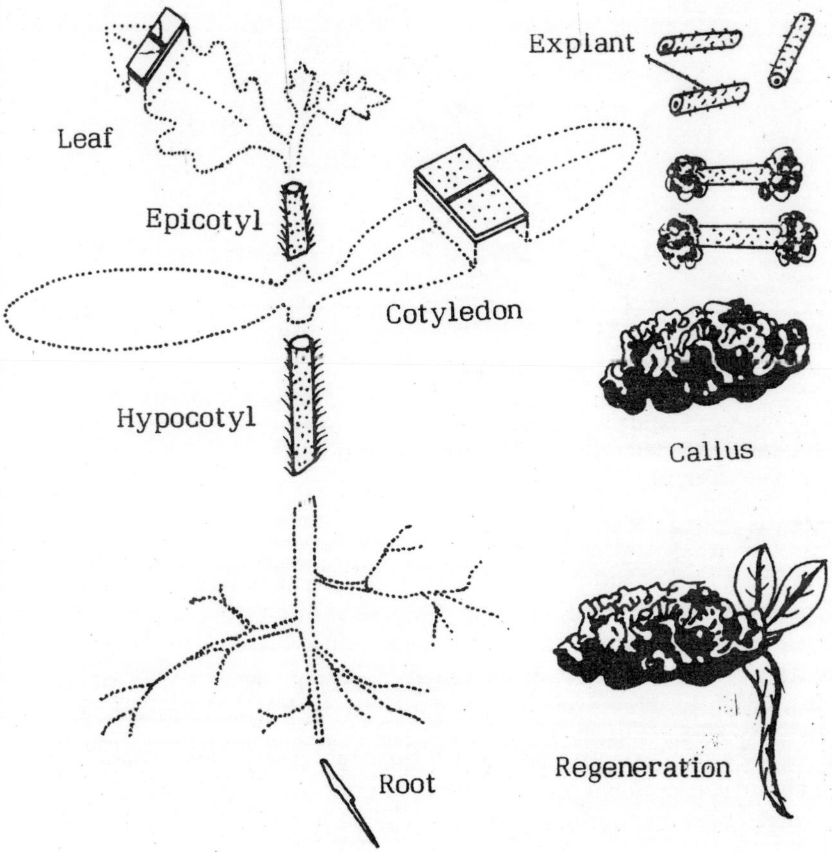

Fig.4.3. Locations of explant sources, callus initiation and regeneration.

CALLUS GROWTH PATTERNS

The growth of callus is measured in terms of increase in fresh weight, dry weight or increase in cell number. A generalized growth pattern takes the form of a sigmoid curve (Fig. 4.4). Three distinct phases can be observed during growth of the callus : (a) the lag phase (b) cell division followed by linear phase and (c) stationary phase and regeneration. Although each stage is characterised by distinct structural and biochemical features but the entire growth cycle is in reality a continuum of physiological changes.

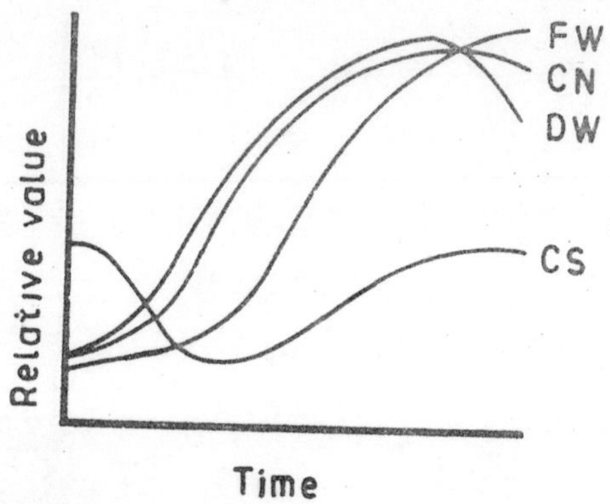

Fig.4.4. Features of a generalized growth pattern of cultured plant cells (CN= cell number, CS= mean cell size, DW= dry weight, FW= fresh weight)

The Lag Phase

This is a period of little or no cell division. When cultured cells are transferred to a fresh nutrient medium, a number of metabolic changes occur. These changes prepare the cells to undergo further division. The levels of NADPH, ATP, cell RNA content and energy charge increase. The rate of total protein synthesis also increases during the lag phase. This increase in protein synthesis is partly due to the efficient ribosomal translation capacity. In *Phaseolus* and *Glycine* subcultures result in rapid formation of polysomes and increase in the translatable levels of group of mRNAs. Presence of auxins 2, 4-D and NAA in the medium increase the mRNA.

During lag phase, only low levels of new compounds are synthesized. An enzyme phenylalanine ammonia-lyase is rapidly synthesised which enhances the activity during lag phase. However, this activity decreases as the culture completes lag phase. Accumulation of complex phenolic compounds do not occur during lag phase.

The Cell Division and Linear Phase

The lag phase is followed by a period of exponential cell division which is followed by a steady rate of division (the linear phase). Once cell division is induced it progresses rapidly. The cell division and expansion are usually confined to the peripheral region of the callus tissue. The number of cells increases which leads to sharp decrease in average cell size. Proliferation is mostly located either in sheet meristems or in nodules at the surface of the callus. When the cell division activity slowsdown, the meristematic activity begins in the deeper tissues of the callus. Chlorophyll formation decreases in rapidly dividing cells. In tobacco, linear phase cells contain poorly developed chloroplasts.

The growth rate of callus tissue can be enhanced by increasing the auxin concentration of the medium. However, the growth rate may be restricted by limiting the levels of essential nutrients in the medium.

The Stationary Phase and Regeneration

As the nutrient supply of the medium depletes, a gradual cessation of cell division occurs. There is sharp decline in the rate of respiration and in DNA synthesis as well as activities of enzymes associated with the pentose phosphate pathway.

Usually the stationary phase is associated with the initiation of structural organisation of the cells which increases the production of secondary metabolites such as alkaloids. Disorganised callus cultures accumulate lesser secondary metabolites. Lindsey and Yeoman (1983) observed that in many Solanaceous taxa at the end of growth phase, organisation and alkaloid accumulation increase together. This supports the view that cellular organisation provides favourable environment for the synthesis and accumulation of secondary metabolites.

The structural and biochemical differentiation result in the regeneration of different organs. Successive subcultures, however, affect the ability of regeneration. In *Hevea brasiliensis* new cultures develop roots spontaneously but this ability of cultures is lost during serial subcultures. Similiarly, single cell clones of tabacco gradually loses the ability to initiate roots and shoots. The potential for secondary metabolite production declines with serial propagation. In cell cultures of *Atropa belladona* accumulation of alkaloid which is dependent upon the formation of root, decreases with subculturing. Prolonged cultures are capable of synthesizing alkaloids. In *Datura innoxia* even after 93 weeks in culture, the callus was capable of synthesizing tropane alkaloids and chlorophyll.

In cultures of immature embroys of *Panicum,* the scutellum produced a compact callus that produced several cup-shaped structures which subsequently differentiated into embryoids. These embryoids germinated when auxin concentration was either reduced in the medium or omitted completely. On the other hand, callus obtained from mature embryos gave rise to shoot buds followed by differentiation and advetitious buds. For induction of embryogenic callus, the developmental stage of the inflorescences where floral primordia have just formed showed best response.

NUTRITIONAL REQUIREMENT

A nutrient medium is needed for the initiation and maintenance of callus cultures. Usually basal medium which consists of a mixture of mineral salts, vitamins, sucrose and phytohormones is used for initiating the callus. The basal medium is supplemented with coconut milk, yeast and malt extracts for additional nutritional requirement of the callus. In forage grasses, supplementation of MS (Murashige and Skoog, 1962) medium with only 2,4-D (2,4 Dichiorophenoxy-acetic-acid) is sufficient to induce callus formations. The continued proliferation and growth of the callus in subsequent subcultures need additional growth hormones or reduced level of 2, 4-D.

For callus induction and proliferation requirement of different growth hormones varies viz., addition of 2, 4-D is essential for callus induction and proliferation. It is not required at the time of organogenesis and plantlet formation. Usually, a high concentration of auxin and low concentration of kinetin promote callus induction and proliferation.

Tissues derived from the green photosynthetic organs may grow in cultures without addition of exogenously supplied sugar. However, for continued growth of callus, presence of exogenously supplied sugar becomes indispensable. Israel and Steward (1966) reported that callus cell in cultures are deficient in some enzymes and cofactors which are essential for synthesis of chlorophyll or proper functioning of chloroplast.

VARIATION IN CALLUS CULTURES

The genetic stability of callus cultures is one of the most important factors that determines the successful utilization of cell cultures for clonal propagation. It has been observed that spontaneous changes occur in the genetic makeup of the prolonged callus cultures. These spontaneous changes may include cytoplasmic structures such as mitochondria and plastids or chromosomes. The cytoplasmic changes may occur either through assymetric cell division or mutation. The chromosomal changes may involve increase in chromosome number, chromosomal breakage, gain, loss or reunion. In callus subcultures fusion of nuclei may takes place which results in the increase of the chromosome number. The rate of these changes vary with the composition of the culture medium, the age of the culture and the species from which explant has been obtained. Some subculture calli do not show any change in the ploidy level even if they are cultured for a period of two years but there are instances where calli show high level of polyploidy within a few months of culturing the explant.Cell suspension prepared from single cell cultures derived from 5 varieties of sugarcane were shown to have variation in chromosome number and karyotype after 6 years in culture.Variation observed among plants regenerated from gametic, somatic and protoplast cultures have been termed respectively as gametoclonal, somatoclonal and protoclonal variation. Somaclonal variation has been attributed to variation already present in the plant material itself and/or to the changes occur during the tissue culture cycle. The latter involves establishment of a more or less dedifferentiated cell mass (e.g. callus), proliferation of a number of cell generations, followed by regeneration of plants.

In *Datura innoxia* plants have been regenerated successfully from aneuploid and polyploid callus without affecting the alkaloid pattern and content of the plants. In tobacco, the plantlets regenerated from the pith of autotetraploid callus accumulated slightly higher level of alkaloid nicotine as compared to the diploid plants of the same species. In callus cultures, therefore, at every stage of cell development, structural and biochemical heterogeneity occur.

Somaclonal Variation

Larkin and Scowcroft (1981) coined the term somaclonal variation to include all those variations which occur in plants regenerated from somatic cultures. Somaclonal variation seems not to be species or organ specific and variation among somaclones have been observed for a wide array of characters. Chaturvedi and Mitra (1975) described two subclones of *Citrus grandis* callus which differed in their morphogenetic pattern. Under identical culture conditions one would consistently form numerous embryoids while the other would form shoots.

Somaclonal variation is a general phenomenon of all plant regeneration systems that involve a callus phase viz., cultured explants, anthers, ovaries and microspores. Somaclonal variation was first reported for vegetative crop plants viz., potato, sugarcane but now this phenomenon is reported for cereals, legumes and other economically important plants (Table 4.1).

Chromosomal Variation

It has been known for sometime that chromosomal abnormalities occur at a high frequency in tissue cultures. In potato (tetraploid with 48 chromosomes) if plants are regenerated from tuber, leaf pieces or stem, nearly 10%

of such plants become aneuploid. Potato plants regenerated from protoplasts the extent of aneuploidy is much greater. It is well documented in sugarcane that chromosome number shifts can occur in cultures and culture derived plants. From a cytological study of sugarcane somaclonal variants, Liu and Chen (1976) concluded that there was no apparent correlation between morphological modifications and changes in chromosome number.

Table 4.1 Some examples of somaclonal variation in economically important species.

Species	Explant	Variant characters
Solanum tuberosum	Protoplast, leaf, callus	Tuber shape, yield, leaf maturity date, plant habit, stem, leaf and flower morphology, early and late blight resistance
Lycopersicon esculentum	Leaf	Male sterility, fruit colour, indeterminate growth
Saccharum officiarum	*Various*	*Eye spot*, *Fiji virus*, downy mildew, culmicolous smut disease, sugar yield, esterase isozyme
Triticum aestivum	Immature embryo	Plant height, spike *shape*, awns, maturity, tillering, α-amylase
Zea mays	Immature embryo	Endosperm and seedling mutants, *D. maydis* race, T toxin resistance, mt-DNA sequence rearrangement
Brassica species	Anthers, embryos, Meristems	Flowering time, growth habit, *Phoma lingam* tolerance

Applications of Somaclonal Variation

Those species which are amenable to cell culture, somaclonal variation may provide an important source of variability for plant improvement. The value of somaclonal variation to plant improvement depends in the ease with which given plant species can be cultured and plants regenerated from cell lines. Somaclonal variation may find its greatest application for plant improvement in concert with selection for desirable mutations at the cellular level. It should no longer be surprising that in the recovery of cell culture mutants the frequencies have been high and mutagenic treatments often failed to noticeably enhance the frequency. Celluler selection is conceivable for the recovery of variants resistant to antimetabolites such as amino acid analogues, antibiotic drugs, pathotoxins, herbicides and physiological stress. Somaclonal variation, cellular selection and early rapid screening of regenerants collectively provide a powerful option for plant improvement.

5

ORGANOGENESIS AND PLANT REGENERATION

ORGANOGENESIS

Organogenesis refers to the process whereby explants, tissues or cell can be induced to form root/and or shoot and even whole plantlets (Fig. 5.1). In other words, the formation of organs is called organogenesis. Organogenesis may be categorised into two groups: rhizogenesis and caulogenesis.

Rhizogenesis:

The process of root formation is called rhizogenesis (Fig. 5.2A). According to Street (1977) root initiation is the type of organogenesis which occurs in cultured tissues most frequently. The root initiation begins after the formation of buds on the cultured tissues.

Caulogenesis:

The process of shoot initiation is known as caulogenesis (Fig. 5.2B). Histological study of shoot organogenesis from embryo derived callus revealed that leaf primordia originated within minute indentations on the surface of the callus. Normal leaf primordia were subsequently formed around the abnormal apices. When transferred to shoot induction medium, organogenesis can be detected within 3 days by radial divisions in the callus. Within 9 days, procambial and provascular tissues are visible in radial tissue. The outermost layer of the callus mass becomes stratified and meristematic. Leaf primordia develop from the surface of cells.

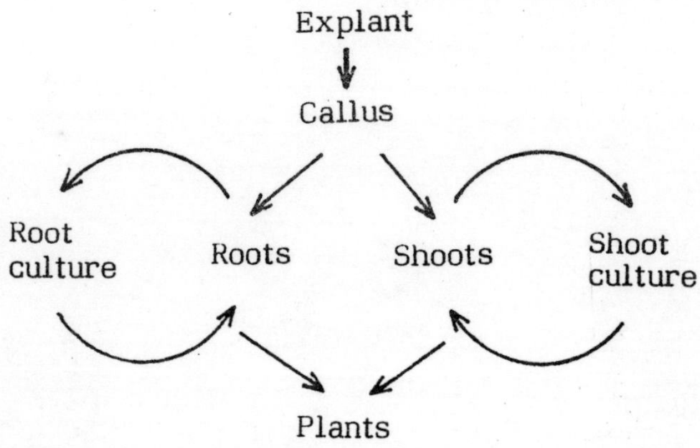

Fig.5.1. Steps in organogenesis.

REGENERATION

Regeneration may be defined as the tendency shown by a developing oraganism to restore any part of it which has been removed or physiologically isolated and thus to produce a complete individual.

In regeneration, mature or nearly mature cells may sometimes become embryogenic again and then undergo changes that restore a disturbed tissue pattern. In this process, there is marked increase in cytoplasmic content and nucleus becomes prominent. This is followed by acceleration of metabolic activity. This phenomena is called dedifferentiation. Dedifferentiated cells become meristematic and resultant tissues may differentiate again in confirmity with its new function or position in the regenerated structure.

In higher plants, regeneration may be of three types: reconstitution, restoration and reproductive regeneration.

1. **Reconstitution**: It includes reorganisation of the embryonic tissue by which its original structure is reformed. This is usually confined to truely embryonic regions such as growing points and young embryos.

2. **Restoration**: It includes the development of missing tissues or organs through the mersitematic activity in the regions adjacent to the place where loss has occurred. During restoration, dormant buds or primordia become active as seen in the formation of new roots and shoots during vegetative reproduction.

3. **Reproductive regeneration**: This is also called vegetative reproduction. During this process, a part of the vegetative body under favourable conditions, gives rise to a new individual.

Fig.5.2. A. Rhizogenesis in leaf callus of *Dioscorea floribunda*. **B.** Regeneration of multiple shoots from stem callus of *Citrus grandis*.

ORGANOGENESIS AND REGENERATION PATTERNS

Plants can be regenerated from cultured materials either by organogenesis-usually the formation of shoots- or by somatic embryogenesis (Fig.5.3). Both processes involve the stable and progressive restriction of developmental potencies. By using organ formation as an end point, it is possible to establish when this determination becomes fixed. Under inductive conditions, tobacco pith tissues only form shoots in regions of the explants closest to the agar substratum (Meins, 1986). Totipotent cells have the potential to form the full range of plant organs. However, under a particular set of experimental conditions, specific types of organogenesis occur excluding other types. This indicates that cultured cells vary in their competence for morphogenesis and responsiveness to inducers. Distinct states of competence for embryogenesis, root formation and shoot formation can be recognised by their stability and conditions for induction.

High frequency plant regeneration is important for cloning, tissue culture breeding and disease eradication. Regeneration can be accomplished either through *in vitro* culture of meristematic explants (isolated meristems, shoot tips and axillary buds) or non meristematic explant (leaf, petiole and root). The difference between the two explant sources is that the meristematic explant consists of cells with a development commitment to the plantlet pattern of development, whereas the nonmeristematic explant cells are committed to alternative developmental pathways. Non meristematic cells can redifferentiate in tissue culture and become committed to the plantlet pattern of development when subjected to an appropriate culture medium.

Formation of localized meristems in cell cultures or induced organogenic-determined cell (IODC) leading to the organization of a well defined shoot and/or root meristem has been reported by Evens *et al.* (1981). The primordia develop in the form of small groups of cells and are destined to form either shoot or root. The redetermination of cells of mature tissue to form root/shoot depends upon the ratio of cytokinin to auxin in the culture medium.

In anther cultures, regeneration of plants takes about 8-12 weeks. The callus obatained by culturing the anthers, when subcultured on MS + NAA (5.4 μm) + BA (8.8 μM) started to differentiate within a week to form shoots. If the differentiating shoots were left on the same medium (having BA 8.8 μM) they turn yellow and finally die. When the callus was transferred to basal medium containing NAA (2.7-5.4 μm) rhizogenesis was initiated in some cases. Morphogenesis may vary in different species of the same genus. In *Arachis villosa*, anthers undergo little proliferation but regenerated shoots

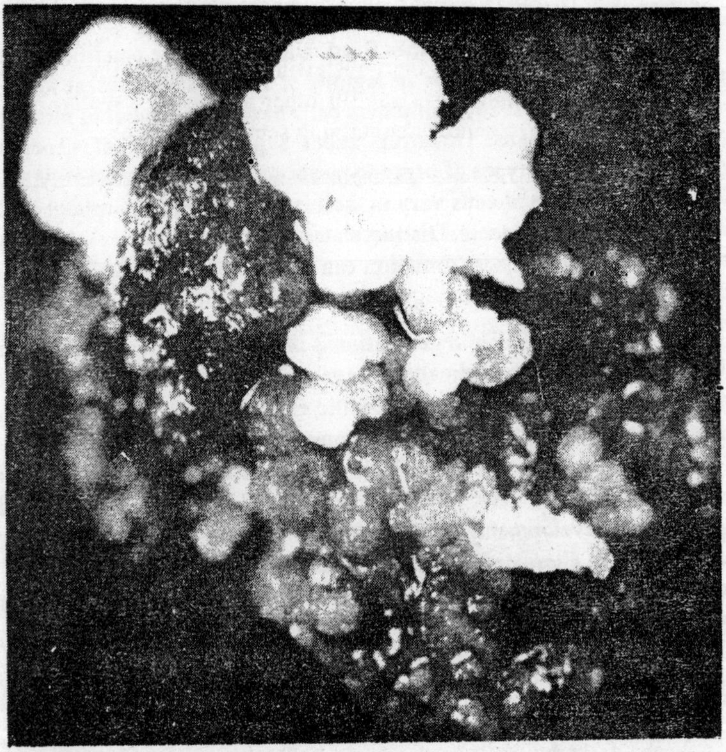

Fig.5.3. Somatic embryogenesis in leaf callus of *Dioscorea floribunda*- embryoids of various developmental stages differentiated at the surface of callus tissue.

and plantlets proliferate frequently. The anthers of *A. hypogaea,* however, proliferated profusely to from a mass of callus which underwent rhizogenesis and occasionally formed shoots.

Tubers provide an excellent explant source for organogenesis in cultures. In *Dioscorea* organogenic or embryogenic cultures, two types of tissues are seen; a pale friable tissue and more compact tissue with small nodules and often roots. It is from the latter that regeneration occurs. Success in regenerating shoots depends upon the transferring and growing the appropriate cell or callus types. This is also true for other monocots. Usually, regeneration occurs in stationary cultures containing semisolid medium.

Table 5.1 lists the *Dioscorea* species that have been regenerated from unorganised callus or suspension cultures. Presence of 2, 4-D in the medium initiates proliferations and maintains unorganised growth during subcultures. Plants regenerated via somatic embryogenesis in *Dioscorea floribunda* showed great variation in vigour, leaf shape and size and in ploidy level, from diploidy to polyploidy to aneuploidy.

A large proportion of the organs differentiated in long term cultures may become morphologically abnormal. Murashige and Nakano (1965) isolated several single cell clones of tobacco. Some of these clones did not form roots and shoots. The clones which initially showed organogenesis, later on showed drastic decrease in this response with the passage of time . Further studies have revealed that totipotent tissues contained diploid and tetraploid cells but reduced stages were associated with a greater frequency of aneuploidy.

Table 5.1 Organogenesis/Embryogenesis in *Dioscorea* cultures (After Ammirato, 1984)

	Species	Explant	Result
1.	*Dioscorea alata* var. purple fleshed	Bulbil segments	Shoots
2.	*D. macrostachya*	Nodal segments	Shoots
3.	*D. composita*	Nodal segments	Shoots
4.	*D. deltoidea*	Seedling tuber	Multiple shoots, embryos
5.	*D. floribunda*	Excised embryos	Embryo, Multiple shoots

The regeneration of whole plants from cultured cells takes place either by shoot bud differentiation or somatic embryogenesis. Shoot bud and embryo differ from one another in the following points :

Shoot bud	Embryo
1. It is a monopolar structure	1. It is a bipolar structure
2. The radicular end is open	2. The radicular end is closed
3. It develops from a group of cells	3. It develops from a single cell
4. Shoot bud has vascular connection with callus tissue or culture explant	4. Embryo has no vascular connection with callus tissue of cultured explant

The epidermal cells of certain tissues may directly differentiate to form shoot buds. For example, shoot bud develops from the epidermal cells of intact hypocotyl in flax *(Linum usitatissimum)*.

Crown gall tumor cells possess capacity for unlimited growth but these cells do not show organogenic differentiation. This happens due to the permanent loss of totipotenciality of parent cells. In some plants, however, the crown gall bacterium *(Agrobacterium tumefaciens)* induces tumor formation. This type of tumor is called teratoma. The cells of the teratomas have the capacity to differentiate shoot buds and leaves. The shoots regenerated from the teratomas are morphologically abnormal.

FACTORS AFFECTING ORGANOGENESIS

1. Growth Regulators

Relative ratio of auxin and cytokinin plays an important role in organogenesis. In tobacco pith tissue cultures high ration of auxin to cytokinin favours root formation; high concentration of cytokinin to auxin favours shoot formation but equal concentrations of auxin and cytokinin the tissue tended to grow in an unorganised manner (Fig. 5.4). This concept of hormonal regulation of organogenesis is now applicable to most plant species.

Thomas and Street (1978) reported that in some cultures auxins promote root formation whereas rhizogenesis is inhibited in others if exogenous auxins are added in the medium. This indicates that conditions necessary for root initiation in one species may not be suitable for other species. Subculturing greatly reduces the capacity of cultures to form roots.

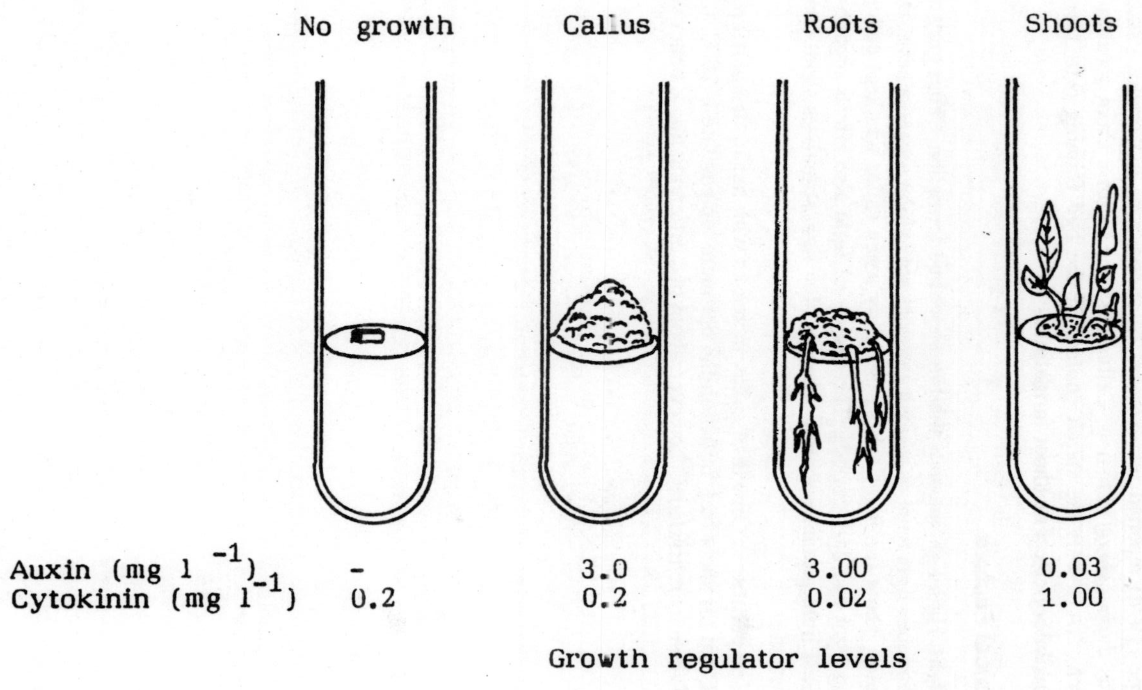

Fig.5.4. Effect of growth regulators on the proliferation and morphogenesis of tobacco pith explant.

In tobacco, bud formation could be suppressed by supplementing the medium with IAA or NAA whereas bud formation may be enhanced by adding purines, adenine and adenosine in the medium.

Thorpe (1978) reported involvement of endogenous gibberellin - like compounds during shoot initiation. Callus destined to form shoot accumulates starch. GA lowers the starch content of the bud forming cells thus causing inhibitory effect on shoot formation.

2. Physical Factors

(i) **Light**: High light intensity inhibits shoot bud formation. Callus grown under continuous light remains whitish and does not exhibit organogenesis. In *Pelargonium* shoot formation takes place only when callus was kept under alternating light and dark periods. The quality of light also affects organogenesis. Red light promotes rooting whereas blue light stimulates shoot bud differentiation.

(ii) **Temperature** : Growth of callus increases with rise in temperature upto 33°C but for shoot bud initiation the optimum temperature is 18°C. In linseed *(Linum usitatissimum),* however, shoot bud differentiation is better at 30°C.

6

SOMATIC EMBRYOGENESIS

The process of embryo development is called embryogenesis. Under certain experimental conditions, the cells of angiosperm sporophyte behave like a zygote and develop into embryo - like structures (Fig. 6.1). Since embryo - like structures are derived from the sporophytic or somatic cells of the plant, the phenomenon is known as somatic embryogenesis. Haccius (1978) defined somatic embryogenesis as "a non-sexual developmental process which produces a bipolar embryo from somatic tissue."

Somatic embryogenesis has been reported in about 71 species of angiosperms distributed over 62 genera belonging to 28 families (Ammirato, 1983). Plants belonging to families Solanaceae, Ranunculaceae, Apiaceae and Rutaceae respond better in producing somatic embryos in culture as compared to other families. Somatic embryogenesis has been induced in a variety of explants viz., stem, leaf, root, hypocotyl, flower, bud, seed, embryo, nucellus and endosperm.

Somatic embryogenesis is of two types; direct embryogenesis and indirect embryogenesis. Direct embryogenesis refers to the development of an embryo directly from the original explant tissue (e.g., nucellar cells of polyembryonic varieties of *Citrus*, epidermal cells of hypocotyl in wild carrot and *Ranunculus scleratus*). Indirect somatic embryogenesis is the formation of embryos from callus or cell suspension, or from cells or groups of cells or somatic embryo (e.g. secondary phloem of domestic carrot, inner

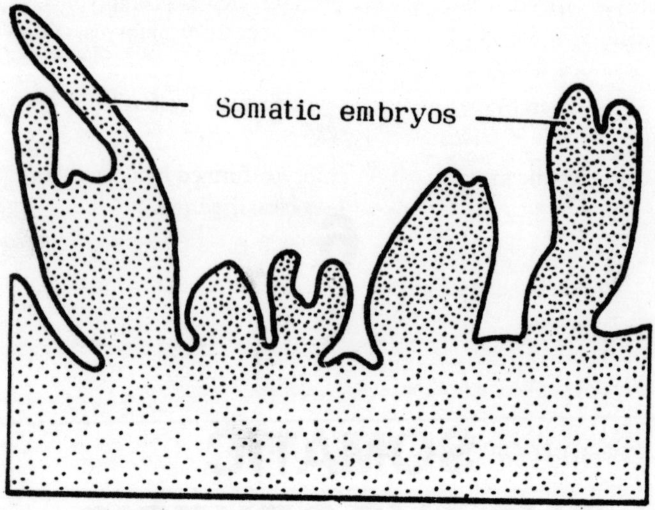

Fig.6.1. Adventive embryoids arising from an in vitro developed plantlet stem (Diagrammatic).

hypocotyl tissues of wild carrot, leaf tissue explants of coffee, pollen of rice). An understanding of these two different patterns of development depends upon consideration of the determinative events of cytodifferentiation during the mitotic cell cycle. The two general patterns of embryogenic development *in vitro* (direct and indirect) may be further characterized by the relative times of determination and differentiation of the cultured cells into embryogenic cells. A comparative account of direct and indirect embryogenesis is given in table 6.1.

Table 6.1. Comparison of general patterns of embryogenic development

	Direct embryogenesis	Indirect embryogenesis
1.	Development of embryo directly from explant tissue.	Callus proliferation occurs prior to embryo development.
2.	Direct embryogenesis proceeds from preembryonic determined cells (PEDC).	Indirect embryogenesis requires the redetermination differentiated cells, callus proliferation, and differentiation of embryogenic determined cells (IEDC).
3.	PEDCs require synthesis of an inducer substance/ removal of an inhibitory substance prior to resumption of mitotic activity and embryo development.	Cells undergoing IEDC differentiation require a mitogenic substance to reenter the mitotic cell cycle and/or exposure to specific concentration of growth regulators.

Embryos formed in cultures have been referred to as embryoids, supernumerary embryos, adventive embryos and accessory embryos. Kohlenbach (1978) classified embryos into following categories :

1. **Zygotic embryos:** The embryos formed by fertilized egg, or the zygote.

2. **Non-zygotic embryos:** The embryos formed by cells other than the zygote. Non-zygotic embryos are further classified into :

   (a) **Somatic embryos:** The embryos formed by the sporophytic cells.

   (b) **Parthenogenetic embryos:** The embryos formed by the unfertilised egg.

   (c) **Androgenetic embryos:** The embryos formed by the male gametophyte (microspores, pollen grains).

DEVELOPMENT OF SOMATIC EMBRYOS

In the callus cultures, embryogenesis is initiated with the differentiation of a single meristematic cell. These meristematic cells are distinguishable from surrounding cells in having dense cytoplasm and prominent nuclei. Further divisions in these cells result in the formation of a group of cells (Fig. 6.2A). Each proembryoid cell possesses the capacity to pass through the same stages of embryogenesis (globular, heart-shaped to torpedo-shaped) as observed in zygotic embryo development.

In *Ranunculus scleratus* somatic tissues proliferate to form a callus from which embryos appear. These embryos develop from the epidermal cells and also deep seated cells of the callus. Embryo differentiation also takes place in suspension cultures raised from these calli. The somatic embryos germinate in situ or when they are planted individually on a fresh semi-solid medium. It is interesting to note that a fresh crop of embryos develop from the stem surfaces of plantlets obtained from somatic embryos.Studies have shown that stem embryos originate from single epidermal cell and follow the same sequence of embryogeny as observed in zygotic embryogeny *in vivo* (Fig. 6.3).

In *Penisetum* and *Panicum* somatic embryogeneis and subsequent plant regeneration in callus cultures (established from immature embryos and young inflorescence segments) have been reported. Immature embryos excised from the inflorescence 7-15 days after pollination were found to be ideal for culture. In 3-4 week old cultures, cup-shaped structures developed at the periphery of the scutellum. When the scutellar callus or small segments of the callus were transferred to a 2, 4-D free medium tube-like structures, reminiscent of coleoptiles, emerged from the centre of the cup-shaped scutel-

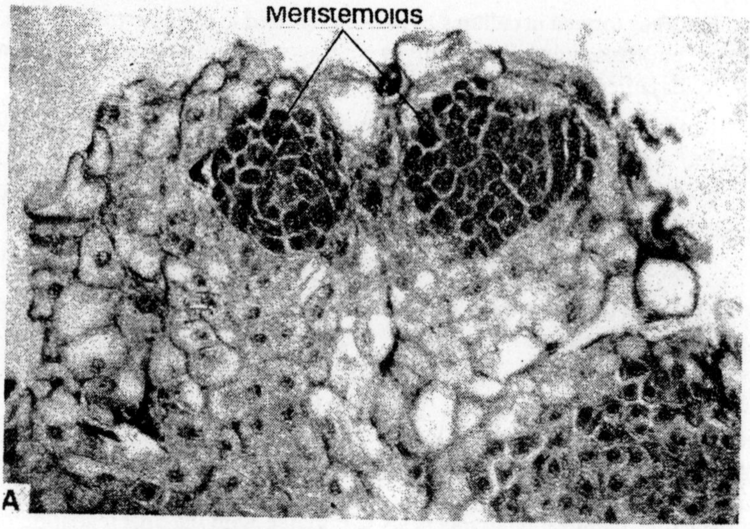

A

B

Fig.6.2. A. Cross-section of proembryogenic tissue of *Rauvolfia serpentina* showing differentiation of two meristemoids- one of the early stages of embryoid differentiation prior to globular stage (A proembryogenic tissue differentiates in callus tissue as discrete patches). B. A sprouting somatic embryoid of *Rauvolfia serpentina*.

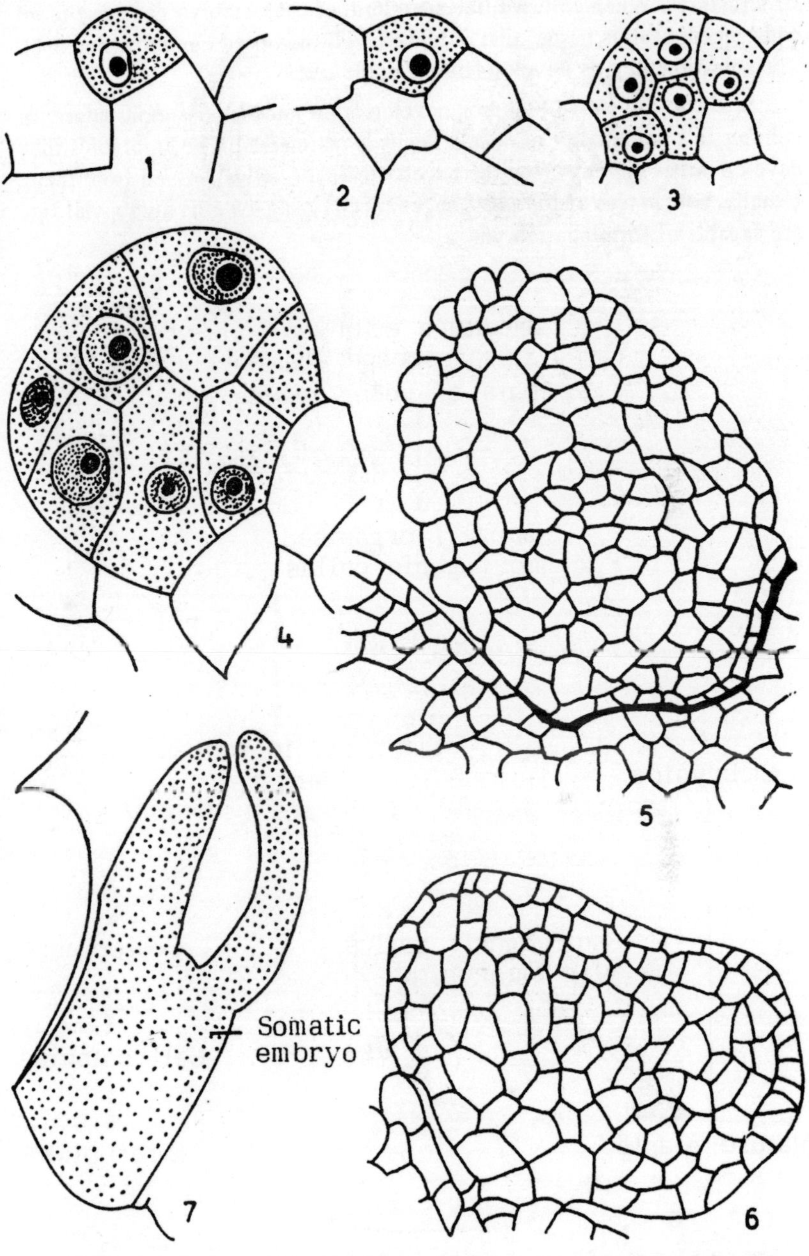

Fig.6.3. Stages in embryo development from epidermal cells of stem of *Ranunculus scleratus*.

lar structures. When cultured, the scutellum of each embryo gave rise to an embryogenic callus tissue that produced well-organised embryos (Fig. 6.4). The somatic embryos develop from a single cell.

In *Citrus* adventive embryos develop from nucellar tissue in nature. In cultures this potentiality of nucellar cells is maintained. The embryoids may develop either directly from the nucellar cells or callus derived from them. Usually, both *in vivo* and *in vitro*, nucellar cells lying towards micropylar side are capable of forming embryos.

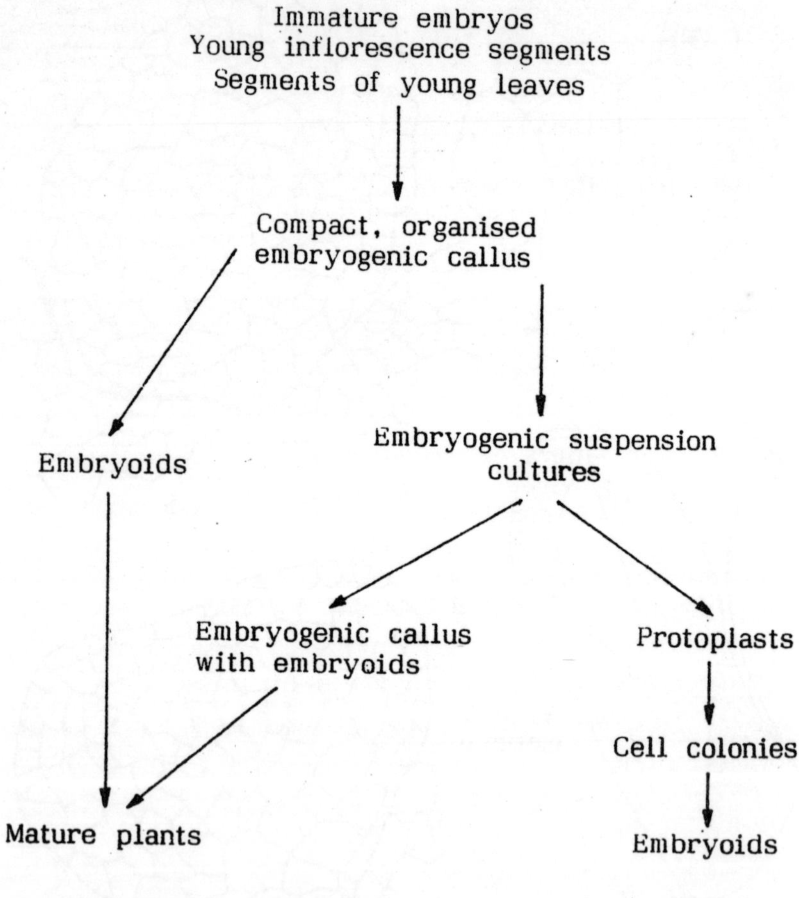

Fig.6.4. Scheme for somatic embryogenesis in *Pennisetum americanum*.

In *Trifolium repens* direct somatic embryogenesis occurred by the proliferatin of hypocotyl epidermal cells of young cotyledonary (torpedo) embryos dissected 8-10 days after pollination on an EC 6 medium containing 0.05 mgl^{-1} BAP (6-benzyl aminopurine). After 3 weeks, embryoids were separated from the parent embryo tissues and rooted on basal medium without BAP. Plantlets were transferred to soil after 1-2 weeks.

Through somatic embryogenesis plants can be produced rapidly and in relatively short periods of 3-6 weeks. Because such plants already possess an integrated shoot-root system, they are easy to establish in soil and grow rapidly into adult plants. It is also possible to synchronize the development of somatic embryos *in vitro*, either by the physical assortment of specific developmental stages, or by the manipulation of culture conditions.

PLANTS FROM SOMATIC EMBRYOS

Somatic embryo can be germinated readily on nutrient medium without hormones until plantlets reach a suitable size for transfer to soil (Fig. 6.2B). From agriculture point of view, somatic embryos must be generated on mass scale. Individual somatic embryos can be encapsulated in an alginate matrix to produce 'artificial seeds'. Such seeds after proper dehydration may be stored for a longer period and may be used when required in bulk for planting in the field. Artificial seeds have been produced in crops such as celery, alfalfa and cauliflower.

CYTOLOGY OF SOMATIC EMBRYOGENESIS

In a proembryogenic mass, cells lying towards the peripheral side are the progenitors of somatic embryos and remaining cells of the proembryogenic mass behave as nurse tissue for the developing embryoids. The ultrastructural studies have revealed that the cells distined to form embryoids contain round, oval to irregular mitochondria, starch filled plastids, a few lipid bodies, free ribosomes and abundant endoplasmic reticulum associated with polysomes. The proembryoid cells have thick wall and they lack plasmodesmata. Till the globular embryo stage the division in the proembryoid cell takes place within the intact cell wall. The wall later on disintegrates to release the embryoids.

GENETICS OF SOMATIC EMBRYOGENESIS

The development of a mature embryoid from a single cell is governed by the genetic and epigenetic mechanisms of the cell. In *Medicago sativa* shoot tip explants obtained from different plants of the same variety show requirement for different levels of auxin and mineral salts. In different cultivars of soybean, the suspensions obtained from hypocotyl or epicotyl explants, under similar cultural conditions, produce variable number of embryoids. Effects of genotype are also very profound in the frequency of occurrence of embryoids having abnormal cotyledons. In *Solanum melongena* the frequency of embryoids developing from leaf explants is influenced by the donor genotype.

FACTORS AFFECTING SOMATIC EMBRYOGENESIS

1. Effect of Growth Substances

The presence of an auxin is essential for somatic embryo formation. If tissue is maintained in auxin free medium, embryos would not be formed. In nucellar culture of *Citrus sinensis* IAA and kinetin were required for its growth and embryo differentiation. In carrot, callus is initiated and proliferated in an auxin rich medium. On such medium, callus differentiates to form localised groups of meristematic cells called embryogenic clumps. When these clumps are transferred to a medium having very low level of auxin or no auxin at all, they form mature embryos.

In *Coffea arabica* somatic embryos are formed when a callus grown on 2, 4-D medium is transferred to a 2, 4-D free medium. The presence of Benzyl amino purine (BAP) in the proliferation medium may promote cell division but it inhibits the embryogenic potential of carrot cultures. Gibberellin inhibits somatic embryogenesis.

2. Nitrogen Source

Somatic embryogenesis is greatly affected by the form of nitrogen source present in the medium. In wild carrot, cultures raised from petiolar segments, embryo formation takes place only when the medium contained some amount of reduced nitrogen. The presence of NH_4^+ in the medium can change the development fate of alfalfa callus cultures from root forming potential. In *Gossypium* embryoids develop only when the suspension is exposed to glutamine.

3. Potassium

The presence of potassium in the medium affects embryogenesis. In domestic carrot, increase in the level of potassium increases the number of embryoids formed. As little as 1 mM potassium salt supports proliferative growth of cells but 20 mM of potassium is needed for optimal embryogenesis.

4. Dissolved Oxygen

High level of dissolved oxygen promotes rooting in the medium but dissolved oxygen below the critical level of 1.5 mgl^{-1} favours embryogenesis.

5. Inhibitors of Embryogenesis

In repeated subcultures and cocultures a marked decrease in the embryogenesis is observed. This is caused due to the acceleration of certain inhibitors. For example, when nucellar tissues of monoembryonic *(Citrus medica)* and polyembryonic (*C. reticulata*) are cocultured, a sharp decrease is noted in the number of embryoids formed.

PRACTICAL APPLICATIONS OF SOMATIC EMBRYOGENESIS

1. Somatic embryogenesis may replace micropropagation for the rapid propagation of economically important plants.
2. Somatic embryogenesis can meet specific breeding objectives by rapidly multiplying germ plasm that is initially present as embryonic material e.g. maternal embryos, haploid embryos and interspecific hybrid embryos that normally abort due to non-availability of endosperm tissue.
3. The nucellus usually degenerates during the development of seed but in *Citrus* species embryogenic callus derived from nucellus remains totipotent for many years and can be used as source material for regeneration of plants.
4. Non-chimeral mutants may be obtained through adventive embryogenesis in tissue cultures as somatic embryos develop from a single cell.
5. Disease free plants can be obtained by using nucellar embryos. In *Citrus* certain viruses which infect the vegetative tissues are eliminated from nucellar cells during ovule development. Somatic embryos developed from nucellar cells produce rejuvenated clones which are therefore, also virus free.

7

EMBRYO CULTURE

The first attempt to grow embryos in a nutritive medium was made by Hannig (1904) who grew mature embryos of *Cochlearia* and *Raphanus* under aseptic conditions. He obtained trasplantable seedlings. Knudson (1922) successfully grew orchid embryos into plantlets by culturing them on agar medium containing sugar. These embryos were cultured without association of a symbiotic fungus which is an essential prerequisite for the germination of orchid seeds in nature. Dietrich (1924) observed that for culture if embryos were excised from mature seeds, and cultured on a solid medium containing Knop's mineral salts and 2.5-5% sucrose, the embryos grew normally but the embryos isolated from immature seeds, skipped stages of normal embryogenesis and directly grew into seedlings. The latter type of growth is designated as "precocious germination". Laibach (1925, 1929) demonstrated that in crosses between *Linum perenne* x *L. austriacum*, the seeds obtained were very light, shrunken and incapable of germination. He could raise the hybrid plants from the embryos dissected out from these seeds.

CULTURE TECHNIQUES

Excised ovules are sterilized and embryos are dissected out and transferred to culture medium. Preferences are given to plants grown under controlled conditions as they provide uniform embryos. In cases where seeds are minute (e. g. orchids) the ovules are cultured and treated as embryo cultures.

In *Capsella bursa-pastoris*, Raghavan and Torrey (1963) employed the following procedure to isolate embryos at different stages of development. Sterile capsules were kept in a few drops of the sterile culture medium. An incision was made in the placental region and the two halves were separated apart with the help of a forcep. By carefully teasing apart the ovular tissues, the entire embryos were separated and cultured (Fig. 7.1).

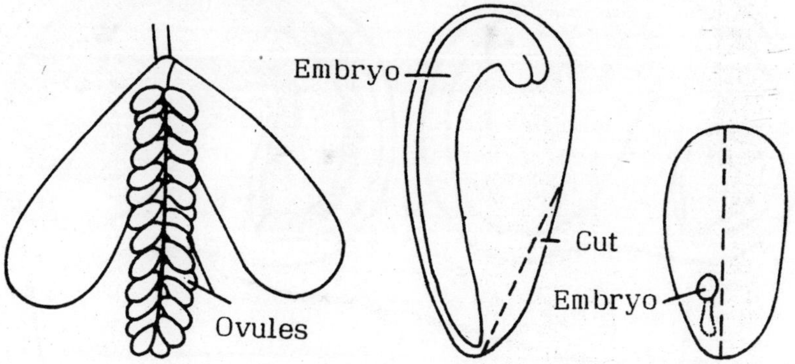

Fig.7.1. Isolation of embryo of *Capsella bursa-pastoris.*

Williams and De Leutour (1980) developed a technique of endosperm transplant for culture of hybrid embryos (Fig. 7.2). The hybrid embryo is removed from its ovule in which endosperm development has failed. In order to provide a transplant nurse endosperm, a normally developing intraspecifically pollinated ovule is dissected to remove the cellular endosperm enveloping a normal embryo. The normal embryo is removed through an exit hole. The hybrid embryo is inserted into the normal endosperm through the exit hole and the endosperm with transplanted hybrid embryo is cultured on the medium.

Culture of Differentiated and Mature Embryos

Mature embryos require an extremely simple medium foı their growth and differentiation. The mature embryo of *Allium cepa* when cultured on NM (Nitsch medium) with 50% sucrose, germinated *in situ* after five days. The embryos of mature coconut grow well on MS medium. The mature embryos of *Amyema migueli* and *A. pendula* cultured on a medium containing IAA and CH, germinated in 3-5 weeks forming a massive holdfast and a pair of plumular leaves.

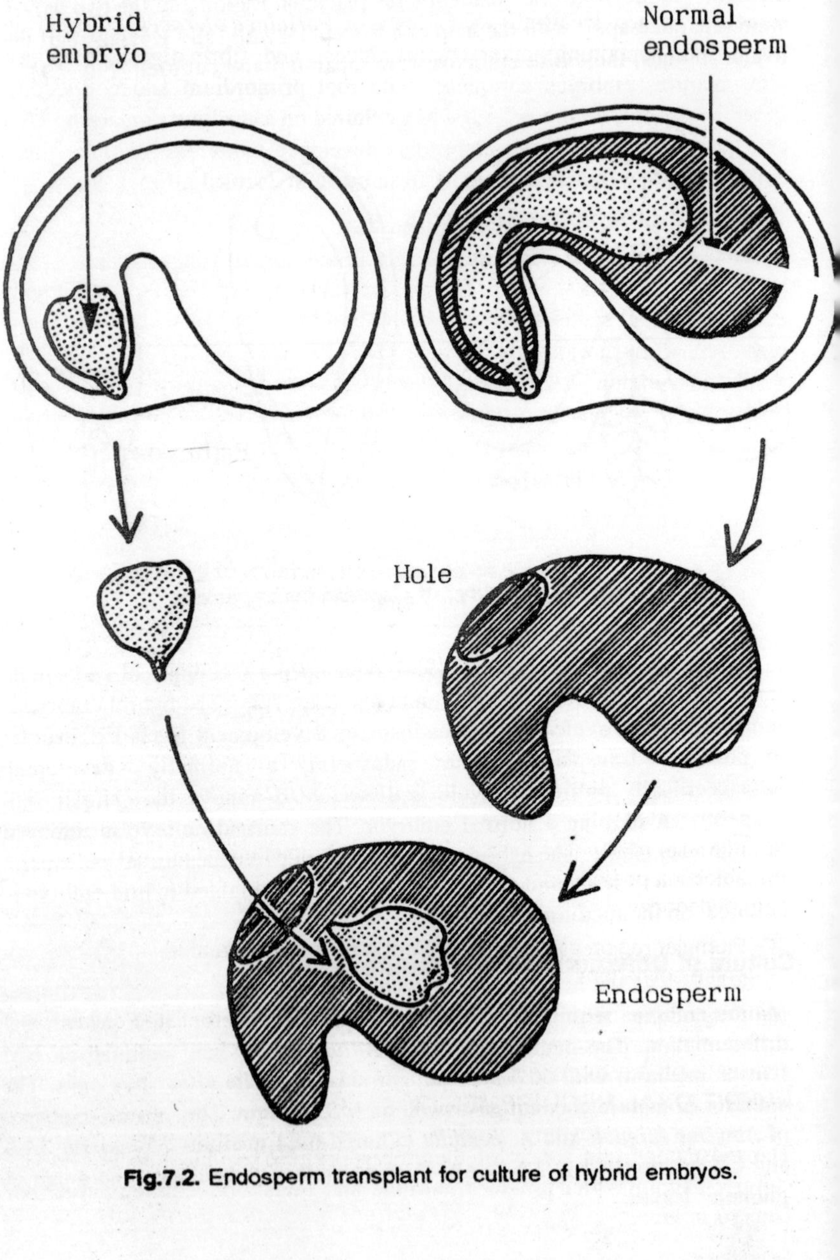

Fig.7.2. Endosperm transplant for culture of hybrid embryos.

Culture of Proembryos

Young and immatue embryos, mostly at heart-shaped or torpedo-shaped stage are cultured. Heart-shaped embryos of *Portulaca oleraçea* cultured on the medium containing mineral salts, glucose and fibrin digest, developed into mature embryos complete with root primordium and cotyledons. Proembryos of *Cuscuta reflexa* when cultured on a medium containing IAA and CH, proliferated and embryoids developed from the callus. These embryoids when transferred on the fresh medium formed normal shoots.

Culture of Embryonal Segments

Cotyledon plays an important role in the normal development of the embryonal axis into a seedling. The mature embryos of maize and other cereals cultured without scutellum failed to attain normal growth of root and shoot primordia. Rangaswamy and Rangan (1971) studied effect of cotyledons on seedling formation in *Cassytha filiformis* and reported that growth factor(s) for shoot morphogenesis resides in the "radicular-halves" of the cotyledons. *In vitro* studies on various segments of dicotyledonous embryos have revealed that only the explants with an intact plumule are capable of forming seedlings (Table 7.1).

Table 7.1. *In vitro* response of various segments of embryo in *Dendrophthoe falcata* (after Bajaj, 1966)

Explant	Growth response
1. Radicular end only	Formed only holdfast with haustoria
2. Radicular end alongwith plumule and adjoining portion of cotyledons	Formed complete plantlet
3. Plumular region alongwith adjoining portions of cotyledons	Formed complete plantlet
4. Plumular region alongwith entire cotyledons	Formed complete plantlet
5. Cotyledons only	Formed only undifferentiated callus

NUTRITIONAL REQUIREMENTS

The most important aspect of the embryo culture is the selection of the culture medium which provides nutritive and metabolic requirements of the embryo *in vivo*.

The composition of the medium and the preparation of the explant for aseptic culture vary with the plant and the age of the embryo to be cultured. Nutritional requirements of the developing embryos of *Capsella bursa-pastoris* are outlined in table 7.2.

Table 7.2. Progressive embryogenesis in *Capsella bursa-pastoris* (after Raghavan, 1966)

Developmental stage	Length of embryo (μm)	Nutritional requirements
Early globular	20-60	Unknown for embryos smallerthan 40 μm
Late globular	61-80	Basal medium + IAA (0.1 mgl^{-1}) + Kinetin (0.001 mgl^{-1}) + adenine sulphate (0.001 mg^{-1})
Heart-shaped	81-450	Basal medium alone
Torpedo-shaped	451-700	Macronutrient salts + vitamins + 2% sucrose
Walking stick shaped and mature embryos	700 and larger	Macronutrient salts + 20% sucrose

1. Carbohydrate

Carbohydrate is one of the most important ingredients which is required as a source of carbon energy for continued growth of cultured embryos. Sucrose is the most commonly used carbohydrate for embryo culture. Sucrose not only provides energy but also maintains a suitable osmolarity. Mature embryo grows well with 2% sucrose but younger embryos need high level of carbohydrate. 12-18% of sucrose has stimulatory effect on the development of excised globular embryos of *Capsella.* Usually 8-12% sucrose have been found to be adequate for the culture of proembryos (e.g. *Datura, Hordeum.*)

2. Amino Acids and Vitamins

The growth of the embryo may be stimulated by adding amino acids to the culture media. Glutamine has proved to be most effective amino acid for promoting the growth of excised embryos. Casein hydrolysate (CH) is being widely used for the embryo culture e.g. *Citrus.* However, sensitivity of embryos to different concentrations of CH may vary with the species.

3. Mineral Salts

Different formulations of mineral salts have been used for embryo culture. For cultured embryos of jute, nitrate nitrogen is essential. NH_4^+ is essential for the growth of immature embryos of *Datura* and barley.

4. Natural Plant Extracts

The coconut milk (CM; endosperm of coconut) is reported to have varying physiological effects in embryos of different species. In 2 or 3 week old maize embryos, coconut milk depressed the growth of root and shoot systems. CM, however, improved the survival and growth of embryos at early cotyledonary stage. In barley, undifferentiated embryos (about 100 cells) could be successfully cultured by supplementing the medium with 20% coconut milk.

Extracts of dates and bananas, tomato juice and wheat glutene hydrolysate have been found to promote growth of excised embryos of barley. The barley endosperm stimulates the growth of embryos of *Raphanus sativus* and *Capsella bursa-pastoris*. This indicates non specificity of the endosperm in embryo culture. Wheat and barley embryos have been successfully cultured on the medium containing maize endosperm.

5. Growth Hormones

Monnier (1978) suggested that growth regulators should not be added to the culture medium as they produce structural abnormalities. Basal medium supplemented with yeast extract, IAA and GA is reported to be useful in successful culture of young embryos of *Corchorus olitorius*. Low concentrations of auxins promote growth of roots and shoots in pea embryos and in *Avena sativa* but at high concentrations growth rate decline. For successful culture of globular embryos of *Capsella* on a mineral-salt-vitamin medium with only 2% sucrose, addition of IAA, kinetin and adenine sulphate has been found to be essential. It appears that young embryos need a higher milieu of growth hormones in comparison of mature embryos. This indicates that physical control of a high osmotic pressure and the chemical control of hormones are linked.

6. pH of the Medium

The pH of the medium is very critical for organogenic differentiation. In barley the optimum pH required is 4.9. However, in *Datura* the optimal pH for early heart-shaped embryo ranges from 5 to 7.5. Rice embryos grow well at two pH values i.e. 5 and 9.

PRACTICAL APPLICATIONS OF EMBRYO CULTURE

1. Production of Rare Hybrids:

In intergeneric and interspecific crosses, in many cases hybrid embryos fail to develop due to poor or abnormal development of the endosperm. But in such crosses the embryo may be potentially capable of normal growth and differentiation. Hybrid plants can be raised by culturing these hybrid embryos before the onset of abortion.

Interspecific hybrids: The embryo culture technique has been employed for production of interspecific hybrids between *Melilotus officinalis* and *Melilotus alba*. Full hybrids have been obtained in interspecific crosses of rice.

The embryo culture technique has been useful in evolving disease resistant tomato. In the reciprocal cross between *Lycopersicon peruvianum* (wild tomato having pest and disease resistant characters) x *L. esculentum* (cultivated tomato) no viable seeds are formed due to abortion of embryos.The embryo callus culture technique has been followed for the production of hybrids.

Intergeneric hybrids: The hybrid seeds in intergeneric cross *Hordeum jubatum* x *Secale cereale* collapse 13-16 days after fertilization. In these cases hybrid embryos fail to develop due to incompatible endosperm. Embryos excised from 9- 12 day old seeds gave rise to seedlings.

In a cross between *Triticum* x *Secale,* the growth of the embryo is retarded in early stages of development. If the female plants are treated with L- amino-n -caproic acid, normal development of the embryo has been observed. Such embryos when excised and cultured grew into normal seedlings.

Embryo culture methods have also been successfully employed for obtaining intergeneric hybrids between *Hordeum* x *Secale, Hordeum* x *Triticum* and *Triticum* x *Aegilops.*

2. Shortening of Breeding Cycle

Long seed dormancy periods pose problems in breeding programmes. But this problem may be overcome by culturing the excised embryos of dormant seeds. Breeding cycle of the plants may be reduced by using embryo culture technique. Rose plants usually take one year to come to flower. Through embryo culture method this period can be reduced to six months.

Malus seeds when grown in the soil, usually take 9 months to germinate but excised embryos produce seedlings within 5 months from the time of culturing.

3. Breaking of Seed Dormancy

One of the major problems in propagation of many early fruiting horticulture plants is the dormancy of seeds. The only method, other than usual practice of grafting, to get progenies is by dissecting out the embryos and culturing them on the medium.

4. Propagation of Rare Plants

The seeds of *Musa bulbisiana* fail to germinate but seedlings can be obtained by culturing the excised embryos on a suitable culture medium. Similarly, in *Colocasia esculentum*, the sterility barrier may be overcome by culturing the young embryos.

5. Determination of Seed Viability

Germination of excised embryos is regarded as more reliable test for determining the seed viability than usual staining methods. In peach seeds and many cereals, long dormancy periods pose problem in testing the viability. In such cases, embryo culture technique can help in determining the seed viability.

8

MICROPROPAGATION

In nature asexual reproduction takes place either by vegetative means (regeneration of new plants from vegetative parts) or by apomixis (seed development without meiosis and fertilization). The vegetative reproduction produces genetically identical plants and is widely used for the propagation and multiplication of horticulturally important plants. The multiplication of genetically identical copies of a cultivar by asexual means is called **clonal propagation** and a genetically uniform assemblage of individuals derived originally from a single individual by asexual propagation constitutes a **clone**. The word clone (Greek clone = twig, spray or a slip, like those broken off as propagules for multiplication) was first used by Webber (1903) to apply to cultivated plants that were propagated vegetatively. *In vitro* clonal propagation is called **micropropagation**.

Micropropagation has a number of advantages over conventional methods of clonal propagation such as (1) a large number of plants can be produced from a single explant within a short period and space; (2) *in vitro* cultured plants are disease free and (3) micropropagation may be carried out throughout the year irrespective of seasonal variation.

MICROPROPAGATION TECHNIQUES

The process of micropropagation involves four distinct stages:

1. Selection of suitable explants, their sterilization and transfer to a nutrient medium,

2. proliferation of shoots on multiplication medium,
3. transfer of shoots to a rooting medium and
4. transplantation of plants in soil (Figs. 8.1,8.2).

SHOOT MULTIPLICATION

In vitro multiplication of shoots involves three main approaches : (1) multiplication through callus culture, (2) multiplication by adventitious shoots and (3) multiplication by apical and axillary shoots.

1. Multiplication through Callus Culture

A large number of plantlets can be obtained from callus either through shoot-root formation or somatic embryogenesis. Mass production of callus followed by shoot regeneration would be the ideal method for large scale propagation of desired plants but there are two serious draw backs : (1) due to repeated subcultures the capacity of mass calli to regenerate shoots is diminished or even lost, (2) degree of aneuploidy, polyploidy and development of genetically aberrant cells progressively increases resulting in the regeneration of plants that differ from the parent type. Therefore, multiplication of shoots through callusing is less preferred method.

2. Multiplication by Adventitious Shoots

Adventitious shoots are stem and leaf structures which develop naturally on plants at places other than normal leaf axil regions. These structures include stems, tubers, bulbs, corms and rhizomes. In many horticultural plants, vegetative propagation through adventitious bud formation is commercial practice. For example, in nature shoots develop on leaves of *Begonia* and some other ornamental plants. In culture also similar type of adventitious shoot formation can be induced by using appropriate combination of growth regulators in media. Adventitious buds can also be induced on the leaf and stem cuttings of even those species which are normally not propagated vegetatively e.g. flax *(Linum usitatissimum)*, *Brassica* species.

Adventitious buds may be formed from a piece of shoot apex. In grapes, from an excised shoot tip more than 8000 plants can be obtained within 3-4 months. *Pinguicula* is usually propagated by leaf pieces and in conventional methods only one plant is produced from each leaf but in cultures 15 fold increase in propagation has been achieved.

African violet (*Santpaulia ionantha*) is conventionally propagated from leaf cuttings which give rise to one or more shoots developing from the base of each petiole. *In vitro* culture of petiole can yield upto 20,000 plantlets from each petiole. Similarly large scale propagation of *Petunia hybrida* and *Kalanchoe blossfeldiana* may be done using small leaf or stem pieces.

Development of adventitious shoot directly from excised organs is preferred for clonal propagation as compared to callus formation method because in former diploid individuals are formed but in latter often cytologically abnormal plants are produced. However, in varieties which are genetic chimeras, multiplication through adventitious bud formation may lead to splitting the chimeras to pure type plants.

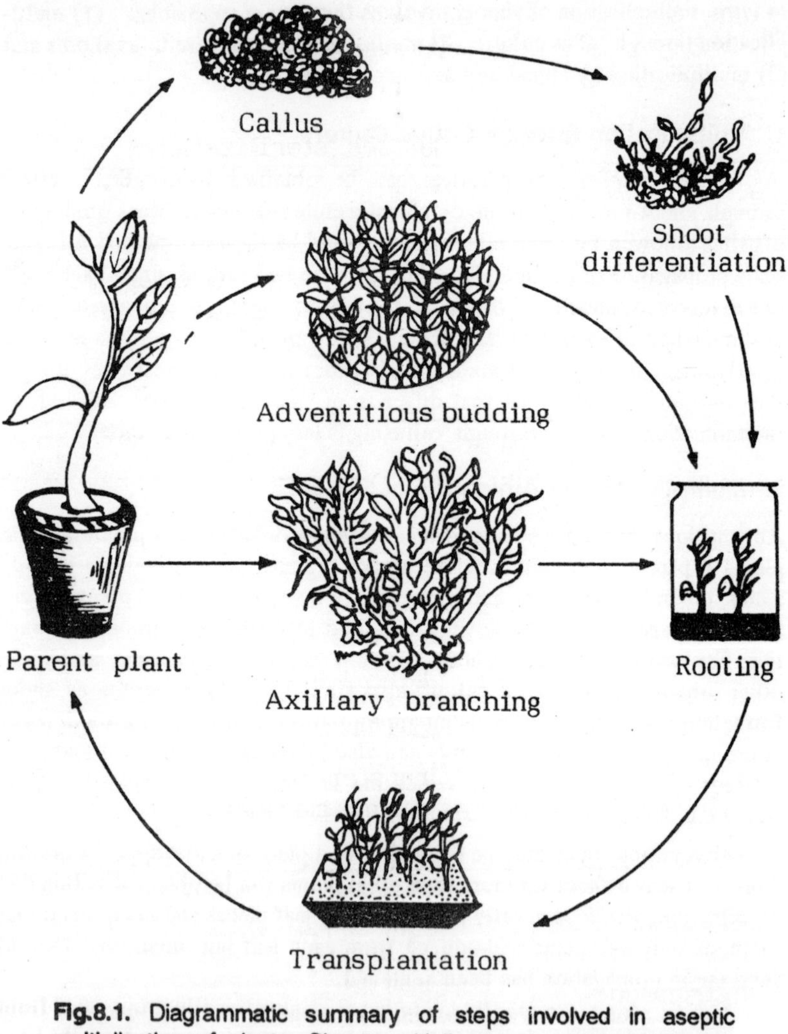

Fig.8.1. Diagrammatic summary of steps involved in aseptic multiplication of plants. Shoot multiplication is achieved through axillary branching, adventitious budding from pieces of plant parts or callusing.

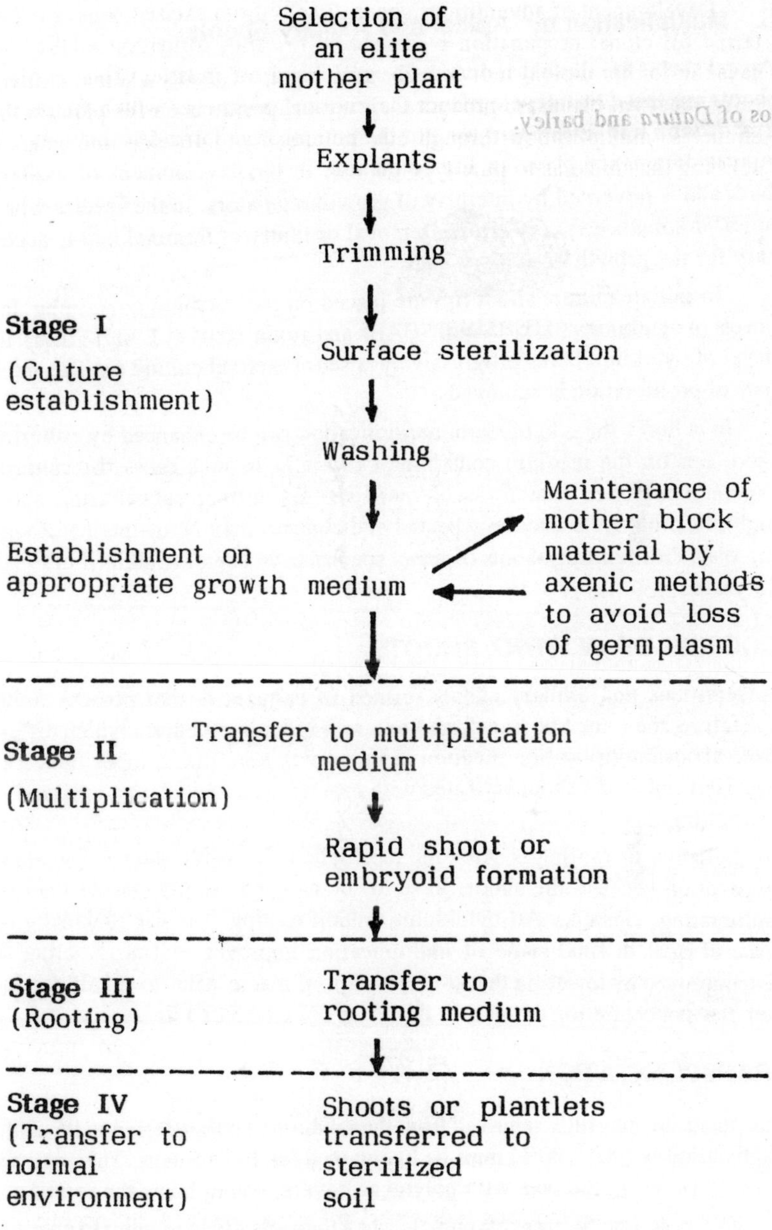

Fig.8.2. Stages of micropropagation.

3. Multiplication by Apical and Axillary Shoots

Apical shoots are those that occupy the growing tip of shoots whereas axillary shoots are those that develop from their normal positions on the plant in the axil of each leaf. Every bud has got the potentiality to develop into a shoot. Apical dominance plays an important role in the development of axillary buds and is governed by interplay of growth regulators. In the species where apical dominance is very strong, removal or injury of terminal bud is necessary for the growth of axillary buds.

To initiate culture shoot tips are placed on the medium containing low levels of cytokinins (0.05-.5 mgl^{-1} BAP) and auxin (0.01-0.1 mgl^{-1} IBA), the level of cytokinin being progressively raised at each subculture until desirable rate of proliferation is achieved.

In cultures the rate of shoot multiplication can be enhanced by culturing shoot tips on the medium containing cytokinin. In such cases the cultured explant transforms into mass of branches. By further subculturing shoot multiplication cycle may be repeated and cultures may be maintained round the year. Proliferating shoots of many species have been maintained upto 10-15 years.

ROOTING OF *IN VITRO* SHOOTS

Adventitious and axillary shoots formed in cultures do not possess roots. Therefore, they must be transferred to a root initiating medium which differs from shoot multiplication medium in its growth hormone composition. The basal cut end of the shoot is treated with root initiating powder and planted in the potting mix.

Initiation of rooting is often the most labour-intensive stage of *in vitro* propagation because the shoots have to be handled singly rather than in proliferating clusters. All cytokinins inhibit rooting. The use of kinetin in place of BAP in final stage of multiplication improves rooting. Rooting is also improved by lowering the concentration of macro salts to a half or less, and the concentration of sucrose from 2 or 3% to 0.5 or 1%.

TRANSPLANTATION

The plants are carefully removed from the medium and transferred to the soil. High humidity (90 - 100%) must be maintained for 10 - 15 days. This may be done by covering the pots with polythene covers having holes for aeration.

The best time for transplanting is when the roots are beginning to emerge out and the developing leaves are becoming photosynthetically self-supporting.

MICROPROPAGATION OF ELITE PLANTS

Tissue culture technique is becoming popular for propagation of a large number of economically important plants. (Fig. 8.3). In orchids, tissue culture is the only commercially feasible method for large scale propagation.

Association of a fungus is a must for orchid seed germination *in vivo* and also the plants obtained through seeds take 4-5 years time to reach flowering stage. In order to overcome these problems, *in vitro* propagation of orchids is the best possible method. Morel (1960) cultured shoot tips of orchid *Cymbidium* and obtained adventitious protocorms. These protocorms by further subculturing may produce four million plants in a year.

Successful micropropagation has been achieved in a number of elite trees such as poplar, willows, *Eucalyptus, Santalum, Phoenix* and *Tectona.* A serious problem associated with the culture of tree species is the browning of the culture medium. This occurs due to the oxidation of phenolic substances which leach out from the cut surfaces of the explants. This problem has been overcome by culturing the explant in liquid medium for a few days or treating the explant with antioxidant solutions before culture.

Many commercial companies are producing millions of plantlets of ornamentals, woody, vegetable and crop species using micropropagation techniques (Table 8.1).

Table 8.1. Some genera micropropagated in *in vitro*.

Vegetable and crop plants	Ornamental plants	Woody plants
Allium	*Anthurium*	*Betula*
Arachis	*Bromeliads*	*Coffea*
Asparagus	*Chrysanthemum*	*Eucalyptus*
Beta	*Cymbidium*	*Hevea*
Brassica	*Dianthus*	*Malus*
Cynara	*Ferns*	*Phoenix*
Fragaria	*Gerbera*	*Pinus*
Glycine	*Gladiolus*	*Populus*
Linum	*Iris*	*Prunus*
Phaseolus	*Narcissus*	*Salix*
Solanum	*Orchids*	*Santalum*
Vigna	*Phlox*	*Tectona*
Zea	*Tulipa*	*Vitis*

Fig.8.3. Micropropagation of guava. A. Multiple shoot formation from the nodal segments of guava on MS medium containing 1.0 mg l^{-1} BAP after 6 weeks. B. Same after 8 weeks of culture. C. Rooting of microshoots on medium containing 0.2 mgl^{-1} each of IBA and NAA with activated charcoal after 3 weeks. D. Different growth stages of regenerants after 3 to 12 weeks of transfer on soil.

FACTORS AFFECTING MICROPROPAGATION

1. Donor Plant and Explant

Physiological condition of the donor plant and explant greatly affect morphogenesis and rate of proliferation in micropropagation systems. Explants which contain large proportion of meristematic tissue are most suitable. The explants obtained from younger parts of the plants give better result than older ones. The behaviour of explants in culture is determined by the stage of the growth of parent plant and season of obtaining the explant.

Most plants can be readily cultured when explants from actively growing plants are used, but others may produce shoots only from plants that are dormant. The bulb scales of *Lilium speciosum* regenerated bulbils freely when taken from plants during spring and autumn periods of growth, but not when taken during summer or winter.

2. Media

The overall micropropagation performance of an explant is directly controlled by the composition of the medium. Higher levels of cytokinins may promote shoot formation but the growth of individual shoots is arrested. In tropical foliage plants, the concentration of phosphate present in the medium affects the size and number of shoots produced in shoot tip cultures. Likewise, the level of potassium in the medium influences the number of somatic embryos formed in wild carrot.

Activated charcoal at about 2% is often included in nutrient media, its action being to absorb aromatic compounds. Charcoal prevents browning in cultured tissues and stimulates embryogenesis and rooting.

3. Light

Light plays an important role in inducing morphogenesis in cultured tissue. The most suitable light intensity for micropropagation is 1000 lux. Low light intensities result in green and taller shoots while high light intensity induces better rooting.

More adventitious shoots were regenerated on *Allium cepa* tissues cultured in 16-hour day than 8-hour day. With shoot cultures of *Solanum tuberosum* the greatest rate of node formation occurred in continuous light, fewer nodes and weaker shoots being formed in 8-hour day.

4. Temperature

Tissue cultures are usually maintained at 25°C. However, some plant species require varied temperature range for better initiation of shoot and roots. In *Asparagus* shoot tip cultures grew best at 27°C whereas petiole segments of *Begonia* and *Cheimantha* cross produced greatest number of shoots at 18°C.

Tropical species require higher temperature, the optimum for date palm being 27°C and for *Monstera deliciosa* 30°C.

APPLICATIONS OF MICROPROPAGATION

1. *In vitro* propagation is a powerful and attractive tool for the rapid cloning of desirable plants. Newly selected cultivars can be bulked up rapidly to the level of several thousands in months rather than years. Additionally, in tissue cultures plant multiplication can continue throughout the year irrespective of the season.
2. Some heterozygous horticultural crops such as *Gerbera* and *Anthurium* that are usually propagated only by seeds can now be cloned rapidly from individuals selected for vigour and colour.
3. Majority of vegetatively propagated crops are susceptible to virus infection. The advantageous use of micropropagation here lies in the rapid production of disease free material. The technique of "meristem tip culture" combined with heat treatment of the mother plant, is now the established means of virus elimination from infected stock.
4. In some crops production of hybrid seeds is very expensive. Production cost of such seeds may be reduced by *in vitro* multiplication of selected parents. Broccoli, cabbage and cauliflower are examples of plants that can now be propagated *in vitro* and desired parents multiplied for the direct production of hybrid seed.
5. Micropropagation has great value as a potential system of germ plasm storage.
6. Tissue cultures can be used to minimise the growing space in commercial nurseries for the maintenance of stock plants.

9

ANTHER CULTURE AND PRODUCTION OF HAPLOIDS

Anther culture is a means of obtaining haploids from pollen grains. As these haploids contain a single set of chromosomes, they are of special interest to geneticists and plant breeders for developing new crop cultivars. Under appropriate cultural conditions, the microspores within the anthers of some species can be induced to undergo embryogenesis resulting in the formation of embryos from which plants can be regenerated (Fig. 9.1). Haploids have been obtained from anther cultures of a number of species but taxa belonging to families Solanaceae, Poaceae and Brassicaceae respond better in cultures. Pollen plants have also been successfully induced in several woody genera such as *Citrus, Vitis, Hevea* etc.

The first pollen-derived haploid callus was obtained in anther cultures of *Tradescantia reflexa* by Yamada *et al.* (1963) but the callus failed to undergo organogenesis. The induction of haploid embryos from pollen grains of *Datura innoxia* by Guha and Maheshwari (1964, 1966) paved the way for production of androgenic haploids. Bourgin and Nitsch (1967) were the first to obtain mature flowering haploid plants of pollen origin in *Nicotiana tabacum* and *N. sylvestris.* Haploid production through anther and pollen culture has been reported in more than 125 species (Table 9.1). Even in those species where *in vitro* androgenesis has been reported, often the frequency with which haploids are formed is very low.

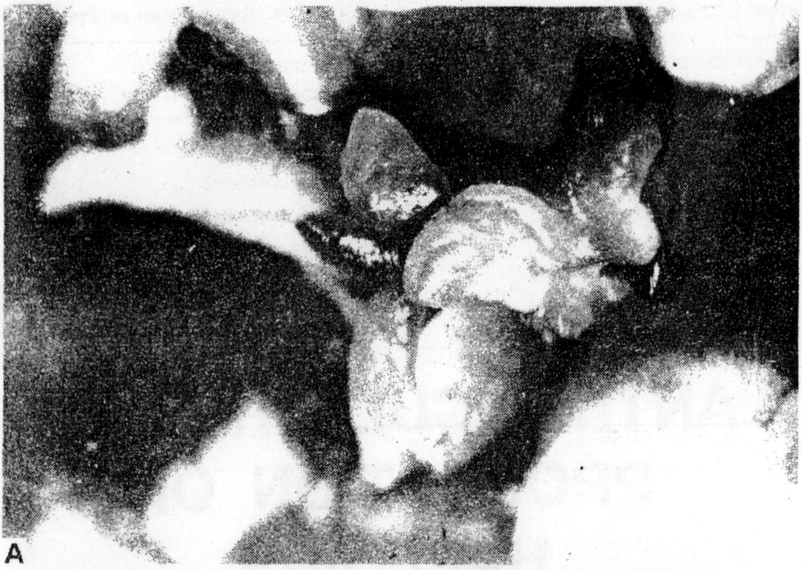

A

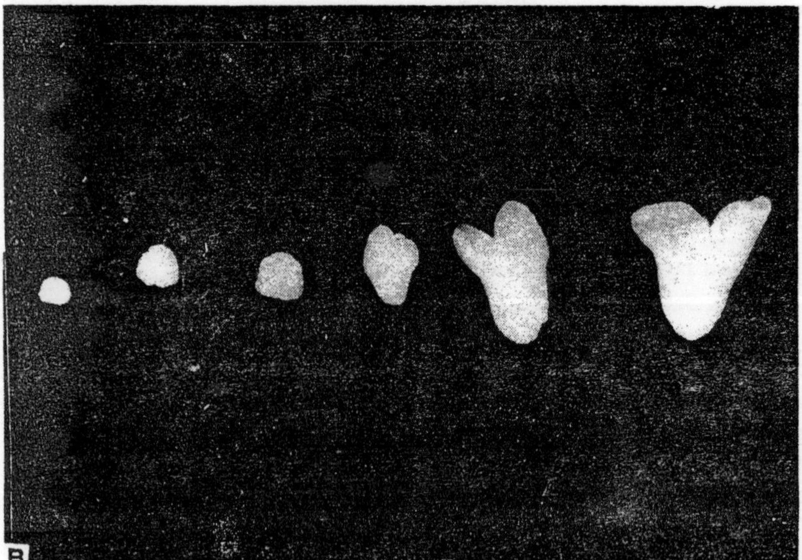

B

Fig.9.1. Androgenesis in *Citrus aurantifolia*. **A.** Differentiation of multiple embryoids produced from within an anther. **B.** Various stages of embryoids isolated from embryoids differentiated in anther.

Table 9.1. Species for which androgenic haploids have been raised

Haploid plants through pollen embryogenesis	Haploid plants through pollen callusing
Brassica campestris	*Arabidopsis thaliana*
Saintpaulia ionantha	*Brassica oleracea*
Oryza sativa	*Hordeum vulgare*
Triticum aestivum	*Triticum aestivum*
Zea mays	*Zea mays*
Atropa belladona	*Asparagus officinalis*
Datura innoxia	*Digitalis purpurea*
Hyoscyamus niger	*Petunia hybrida*
Nicotiana tabacum	*Solanum melongena*
Solanum tuberosum	*S. tuberosum*
Paonia hybrida	*Lolium multiflorum*
Nicotiana langsdorffii	*Lolium longiflorum*

CULTURE TECHNIQUES

Young flower buds at appropriate stage of microspore development are surface sterilized with alcohol or chlorine water and the anthers are carefully removed and cultured on nutrient medium. For culture, both solid and liquid media are used. In anther cultures, depending upon the culture medium, age of anther, age of donor plant and temperature, pollen grains may either directly develop into embryoids or it may first form a callus from which the plantlets would subsequently develop (Fig. 9.2).

Special care should be taken to ensure that the anthers are not injured during operation. Injured anthers should be discarded as they may stimulate undesired callusing. In cases where buds are small and dissection is difficult entire bud is cultured. Sunderland and Dunwell (1977) reported that presence of other floral parts does not affect the pollen response. After culturing, anthers burst open within 3-6 weeks giving rise to pollen callus. Shoots developed from the callus are separated and are transferred to a medium which would support better development of root-shoot system. The plantlets are carefully removed from the medium and planted in the potting mix and finally transferred to the soil.

Anther cultures are usually maintained in alternative periods of light (12-8 hr; 5,000-10,000 lux m^2) at 28°C and darkness (6-12 hr) at 22°C. However, optimal culture conditions vary from species to species.

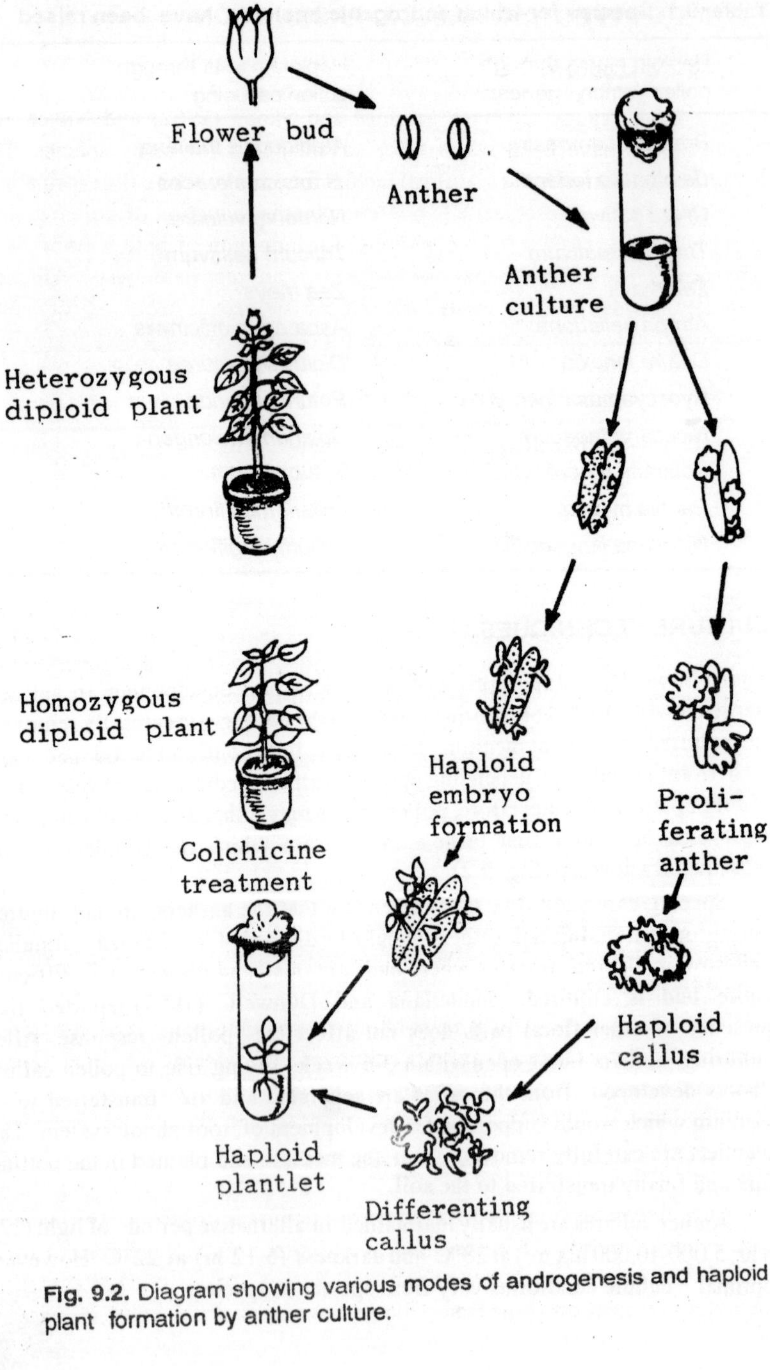

Fig. 9.2. Diagram showing various modes of androgenesis and haploid plant formation by anther culture.

FACTORS AFFECTING HAPLOID PRODUCTION

1. Culture Medium

Culture media suggested by Murashige and Skoog (1962) and Nitsch and Nitsch (1969) have been found to be advantageous for many species. The hormones in the medium are critical factors for embryoid or callus formation. In cereals, auxin, particularly 2, 4-Dpromotesthe induction of pollen callus. Kinetin or other cytokinins are essential for induction of pollen embryos in Solanaceae. The concentration of sucrose plays an important role in induction of pollen plants. In tobacco and rice, 2-5% sucrose is required for callus formation but 6-12% sucrose is suitable for cereals, rapeseed and potato.

2. Physiological State of Donor Plant

Age and growth conditions of the donor plants affect the embryogenic potential of the cultured anthers. Usually anthers from young plants give better response during androgenesis. In tobacco, highest level of embryogenesis is observed when donor plants are grown under high light intensity and short days. Anther donor plants if subjected to nitrogen starvation result in increase in embryo yield.

Anthers derived from older inflorescences show decline in the frequency of haploid production due to reduced pollen viability. In *Datura innoxia* plants grown at 20-30°C gave better anther response. In barley a high frequency of equal divisions in pollen nuclei and callus formation could be achieved if plants are clipped at the ground level and allowed to stand in water for 1 or 2 days before culturing the spike.

3. Genotype of Donor Plants

The genotypic differences among the donor plants greatly affect the ability of pollen grains to form haploid plants e.g. in *Nicotiana* most of the species produce pollen plants readily but in *N. langsdorffii* only a few pollen embryos could be induced.

By intercrossing poorly responding clones androgenic response can be improved. This technique has been employed in *Solanum tuberosum* (Jacobsen and Sopory, 1978) and *Hordeum vulgare* (Foroughi-Wehr *et al.* 1982) and the lines developed showed better response than either of the parents.

4. Developmental Stage of Pollen

The developmental stage of microspores at culture is critical for the induction of division. Usually the uninucleate microspores are more suitable for induction of androgenesis. In *Nicotiana* and *Datura* most responsive stage of embryoid formation is the first pollen mitosis stage.

Anthers of *Datura* and *Petunia* with uninucleate microspores give rise to haploids. Anthers with binucleate microspores give rise to diploids while those with late binucleate microspores to triploids or polyploids.

5. Physical Factors

Temperature and light affect the formation of pollen embryoids. In general, a temperature range between 23°C-25°C is suitable for pollen divisions. Sometimes, pretreatment of anthers at high temperatures may improve the rate of androgenesis. The light also affects pollen embryogeny. In *Lycopersicon* and *Digitalis* and in most cereals, incubation of cultures in darkness favoured callus growth from microspores. However, regeneration of plantlets occurred only when cultures were transferred to light. In *Brassica* maximum number of embryoids were obtained when anthers were cultured for 21 days at 30°C before transfer to 25°C. In *Nicotiana* and rye, cold treatment given to anthers prior to isolation and culture of microspores stimulates embryoid formation.

IN VITRO ANDROGENESIS

Normally during the course of pollen development, the uninucleate microspores undergo an unequal division forming a small generative cell and a large vegetative cell. The generative cell again divides to form two male gametes. In anther cultures a microspore can form sporophyte through any one of the following pathways :

(i) *Pathway-I*: The microspores divide by an equal division and the two daughter cells contribute towards the formation of sporophyte.

(ii) *Pathway-II*: The microspores divide by an unequal division and the sporophyte develops from the vegetative cell. The generative cell either does not divide further or divide once or twice before degeneration.

(iii) *Pathway-III*: The microspores divide by an unequal division forming small generative cell and large vegetative cell. The sporophyte is formed by the generative cell only. The vegetative cell either does not divide further or divides to a limited extent.

(iv) *Pathway-IV*: The microspores divide by an unequal division forming small generative cell and a large vegetative cell. Both these cells contribute to the development of sporophyte (Fig. 9.4.).

Irrespective of the early pattern of divisions, the responsive pollen grains become multicellular and burst open to release cellular mass having an irregular shape. This tissue gradually becomes globular and undergoes normal embryonic differentiation e.g. *Datura, Nicotiana*. By this method, one grain

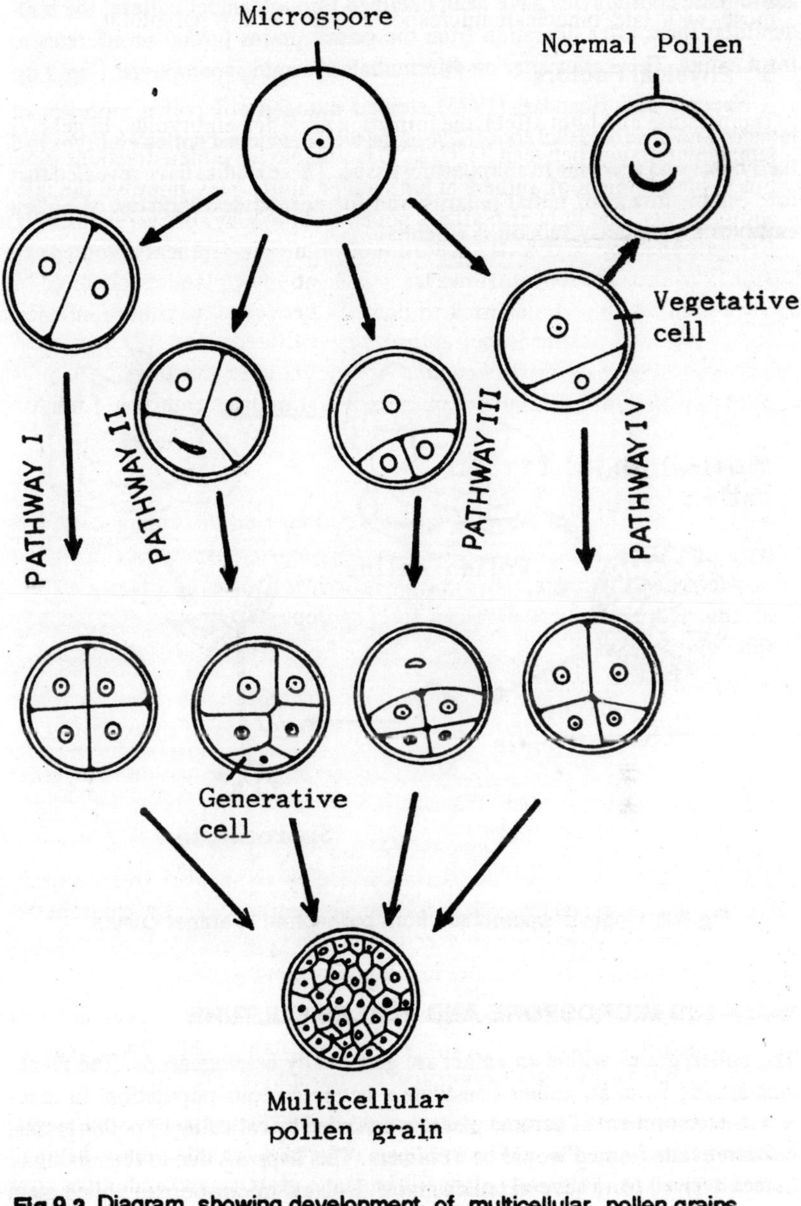

Fig.9.3. Diagram showing development of multicellular pollen grains.

gives rise to one haploid plant. In some plants such as *Asparagus* where androgenic sporophytes have been obtained through anther culture, the multicellular mass after liberation from the pollen grains further proliferates to form callus. This callus later on differentiates forming sporophytes (Fig. 9.4).

Haccius and Bhandari (1975) studied ontogeny of pollen embryos of tobacco and observed that about 82% of the well developed pollen embryos had their basal end adhering to a supporting tissue. These studies have revealed that for establishment of initial polarity and for normal development of pollen embryos a temporary support is essential.

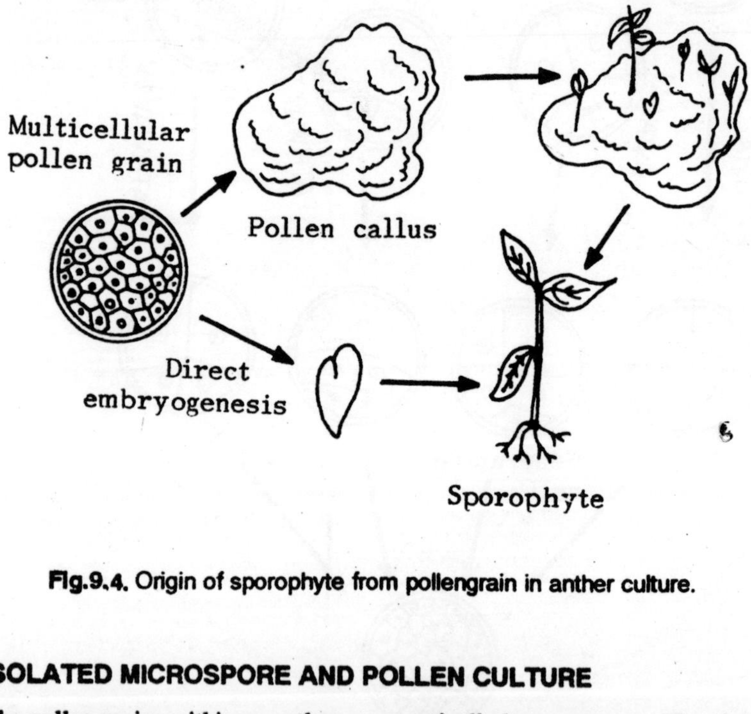

Fig.9.4. Origin of sporophyte from pollengrain in anther culture.

ISOLATED MICROSPORE AND POLLEN CULTURE

The pollen grains within an anther are genetically heterogeneous. The plants thus arising from an anther constitute a heterogeneous population. In cases where development of haploid plants is preceded by callusing of pollen grains, the tissue thus formed would be a chimera. This happens due to the mixing of tissues derived from several pollen grains. Isolated microspore or pollen grain cultures can be a solution to avoid chimera formation. Another problem which can be overcome by isolated pollen culture is the competition between a large number of pollen grains for growth in the limited space available within the anther locule.

Sharp *et al.* (1972) successfully raised haploid tissue clones from isolated pollen grains of tomato through nurse culture technique (Fig.9.5). Without the nurse-anther, pollen did not form colonies. Wenzel *et al.* (1975) successfully demonstrated the possibility of improving androgenic response in pollen culture of *Secale cereale* by enriching the population of androgenic grains naturally occurring through gradient centrifugation.

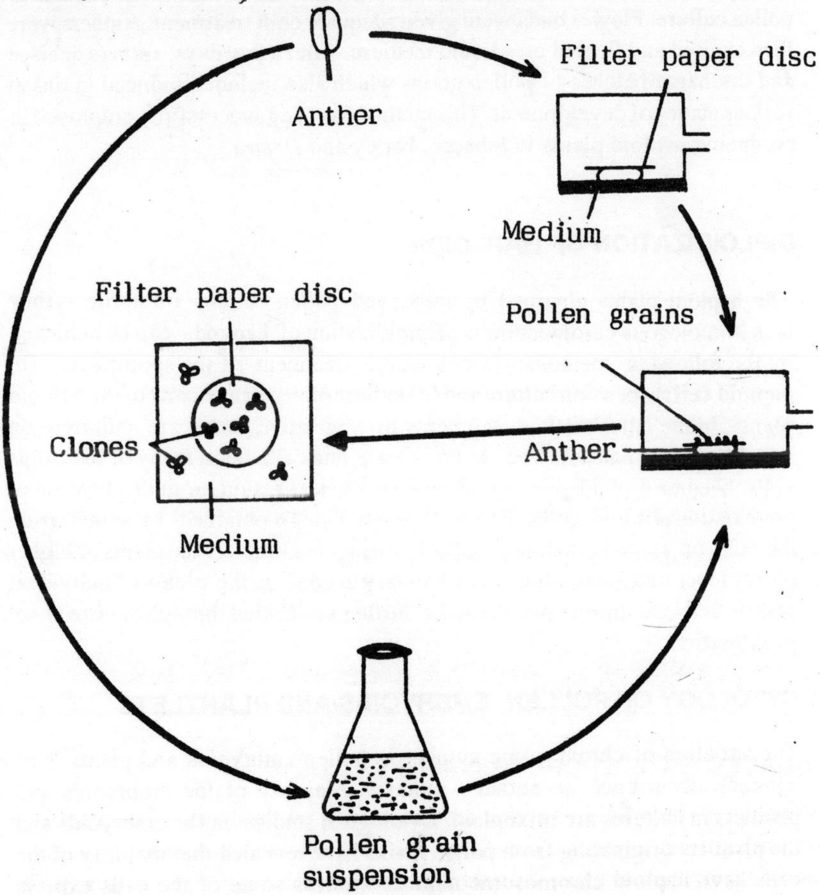

Fig.9.5. Diagram showing the technique of nurse culture for raising tissue clones from isolated pollen grains of tomato.

Nurse Culture Technique

Buds were surface sterilized by hypochlorite solution. Anthers were then removed aseptically and placed horizontally on the top of the basal medium. A filter paper disc was placed over the anther. Suspension of pollen from anthers of another bud was prepared in liquid medium (10 pollen grains in 0.5 ml of the medium) and about 10 pollen grains were placed on the filter paper. After about two weeks, clusters of green parenchymatous cells forming haploid clones appeared on the disc.

Sunderland and Roberts (1977, 1979) proposed a modified method of pollen culture. Flower buds were given adequate chill treatment. Anthers were then excised and floated on a liquid medium. After a few days, anthers dehisce and discharge batches of pollen grains which also included induced grains at various stages of development. This method has been successfully employed in producing haploid plants in tobacco, barley and *Datura*.

DIPLOIDIZATION OF HAPLOIDS

The haploid plants obtained by anther and pollen culture are sterile as they lack homologous chromosomes. Diploidization of haploids can be achieved by the following methods: (a)colchicine treatment of the sporophyte, (b) haploid cell suspension culture and (3) culture of vegetative parts of the haploid plants. In the latter method, segments of vegetative parts are cultured on suitable media and allowed to proliferate into calli. Soon many of the callus cells become diploid due to endomitosis/nuclear fusion or multiplication of pre-existing diploid cells. Fertile diploids can be obtained by transferring the calli on suitable media (Fig.9.6). In order to eliminate the plants of higher ploidy level and aneupolids, it is necessary to confirm the ploidy of individual plants. Selected diploid plants can be further multiplied through *in vitro* shoot proliferation.

CYTOLOGY OF POLLEN EMBRYOIDS AND PLANTLETS

The variation of chromosome number in pollen embryoids and plants is of common occurrence in anther cultures. Majority of the embryoids and plantlets in cultures are mixoploid. Cytological studies on the embryoids and the plantlets originating from pollen grains have revealed that majority of the cells have haploid chromosome number whereas some of the cells express aneuploidy and polyploidy.

Cytological studies of the apical cone of pollen trees of poplar show five types of somatic cells in each pollen tree: (1) diploid cells and aneuploid cells with a chromosome number close to 2n, (2) haploid cells and aneuploid cells with a chromosome number close to n, (3) aneuploid cells with chromosome

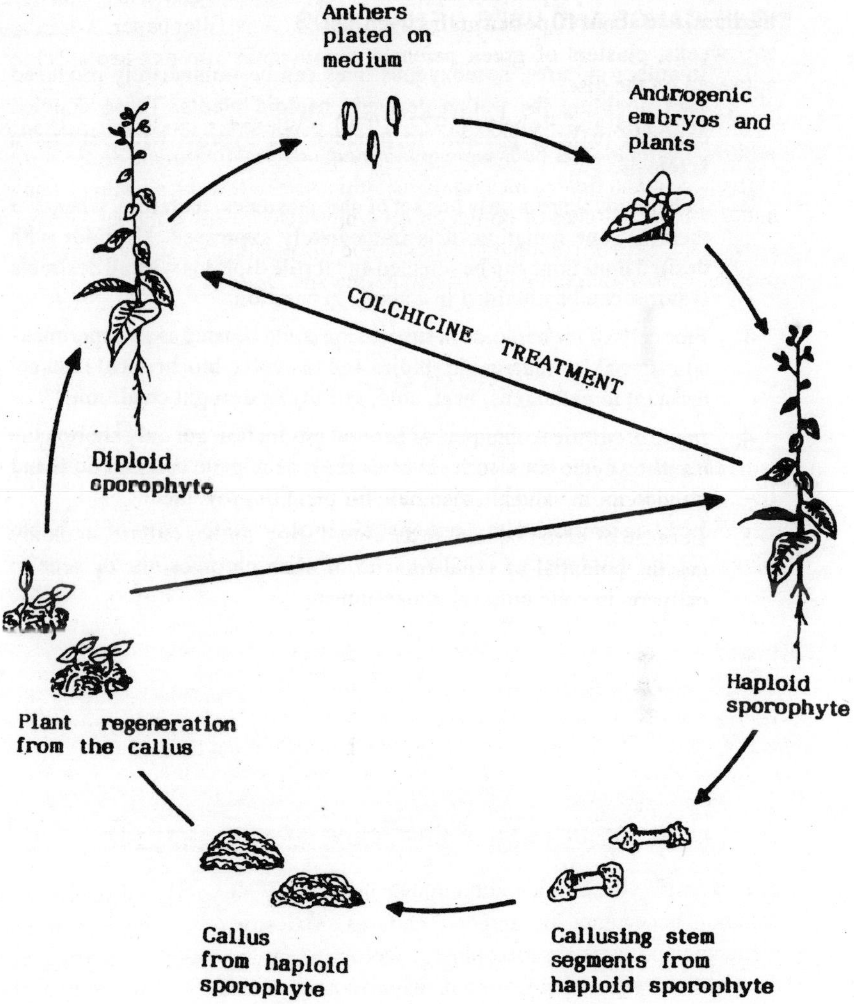

Fig.9.6. Diploidization of anther derived haploids.

numbers between diploid and haploid, (4) polyploid cells and (5) cells with a chromosome number less than n. Some pollen trees grew vigorously as majority of their cells had the diploid chromosome number whereas others grew slowly because the majority of their cells were aneuploid.

SIGNIFICANCE AND USES OF HAPLOIDS

1. In anther cultures, homozygous lines can be immediately produced after doubling the pollen derived haploid plants. These doubled haploids make it possible to produce the hybrids with maximum heterosis.
2. In haploids there is only one set of chromosomes, therefore, whenever there is gene mutation, it is immediately expressed. Haploids with desired mutations can be selected and fertile diploids with all desirable features can be obtained in a single generation.
3. Free cells of the haploids in suspension could be used as an experimental material for mutagenic studies and to evolve biochemical mutants resistant to pathogens, heat, cold, salinity or drought conditions.
4. *In vitro* culture techniques of haploid production not only shorten the breeding cycle but also recover more types of gene combinations and provide more valuable materials for plant improvement.
5. Because of chromosomal instability during anther culture, haploidy has the potential of rapid transfer of alien chromosome or gene to cultivars in wide crossing programmes.

10

PRODUCTION OF SECONDARY COMPOUNDS

From the prehistoric times, man has used the plants as a repository of indispensible raw materials and drugs. Plants synthesize a wide range of chemical substances, many of which possess commercially useful compounds (Table 10.1). Many of these compounds come under the category of secondary metabolites e.g. alkaloids, volatile oils etc. These metabolites may be defined as those substances which are not essential for the survival of the plant. Some plant species have the capabilities to synthesize certain compounds unique to them, in addition to the reactions of the primary metabolic pathways common to all life forms. These reaction sequences are called secondary metabolism and the products of these reactions are known as secondary metabolites. Many of the secondary metabolites are chemically complex and in many instances synthesised substitutes are not effective as the natural substances. Natural products from plants used as pharmaceutical agents show great diversity in both chemical structure (Fig. 10.1) and pharmacological activity (Table 10.1). Plants also provide useful compoumds such as perfumes, flavouring and colouring agents and food additives. Besides these, plants have become an important source of major insecticidal products.

The biotechnological importance of secondary metabolites lies in the fact that : (i) many of them are commercially-valuable chemicals, (ii) a number of them are toxic and should be removed from food products and (iii) they act as protective agents, used by the plant against insects, pathogens or animal foragers. The application of secondary compounds falls into five main categories viz., durgs, flavours, perfumes, pigments and agrochemicals.

TABLE 10.1. The ten most prescribed medicinals from plant resources (after Fowler, 1983)

Medicinal agent	*Activity*	*Plant source*
Steroids from Diosgenin	Antifertility agents	*Dioscorea deltoidea*
Codein	Analgesic	*Papaver somniferum*
Atropine	Anticholinergic	*Atropa belladona*
Reserpine	Antihypertensive	*Rauvolfia serpentina*
Hyoscyamine	Anticholinergic	*Hyoscyamus niger*
Digoxin	Cardiatonic	*Digitalis lanata*
Scopolamine	Anticholinergic	*Datura metel*
Digitoxin	Cardiovascular	*Digitalis purpurea*
Pilocarpine	Cholinergic	*Pilocarpus abonandi*
Quinidine	Antimalarial	Cinchona ledgeriana

PLANT CELL CULTURE AND PRODUCTION OF SECONDARY COMPOUNDS

Plant cells have the potential to produce, either by *de novo* synthesis or by biotransformation of specific precursors, an extensive range of secondary metabolites in culture (Table 10.2). In alkaloid biosynthesis, cultures derived from explants of various parts of an alkaloid producing plant, are capable of synthesizing alkaloids identical to that formed in the whole plant. Some cultures may even produce high level of alkaloids. For example, in nature a number of *Datura* species contain alkaloid aposcopolamine only in traces but this alkaloid predominates in culture extracts.

Callus tissue differentiation plays an important role in the synthesis of secondary compounds. For example, root differentiating calli of *Atropa belladona* produce tropane alkaloids whereas non-differentiated calli do not produce tropane. Likewise, differentiated cultures of *Papaver* produce more thebaine as compared to undifferentiated callus. On the other hand, undifferentiated cultures of *Dioscorea deltoidea* produce 1-2% steroidal sapogenins but organogenesis results in the decrease of sapogenins to traces.

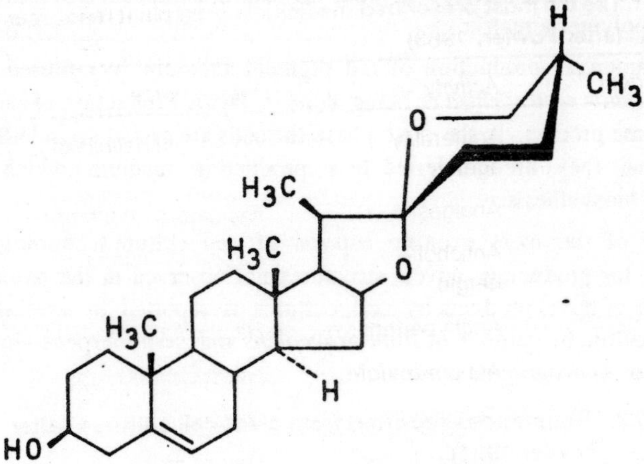

Diosgenin

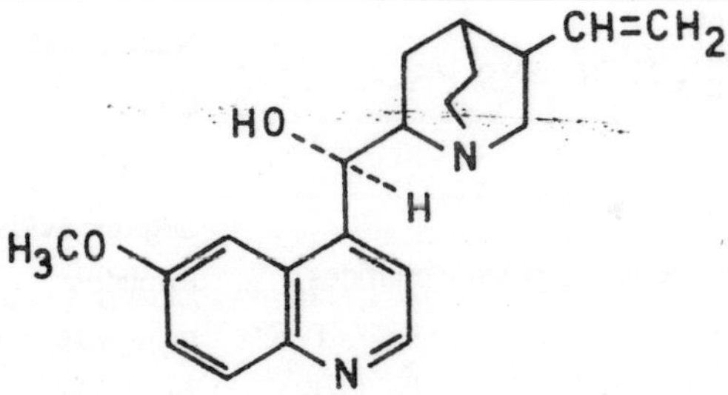

Quinine

Fig.10.1. Chemical structure of commonly used plant products, Diosgenin and quinine.

In cell cultures increased yield of natural products has been reported (Table 10.3).

Large scale production of secondary metabolites involves two main systems i.e. fermenter systems and immobilized cell systems. In the former freely-suspended cells are grown upto a stationary phase and then harvested to extract the product whereas in latter the cells are embedded or entrapped in an inert polymeric matrix.

Commercial production of red pigment shikonin by cultured cells of *Lithospermum erythrorhiza* is being done in Japan. Production of shikonin is a two phase process. In the first phase the cells are grown up in bulk and in the second they are transferred to a production medium which favours shikonin biosynthesis.

One of the most exciting aspects of cell culture technology is the potential for producing novel structures not observed in the parent plant. Synthesis of novel products by cell cultures is reported in several plants, e.g. rutacultin by cultures of *Ruta graveolens* and sesquiterpene lactones by cultures of *Andrographis paniculata*.

Table 10.2. Substances reported from plant-cell cultures (after Fowler, 1981)

Alkaloids	Latex
Allergens	Lipids
Anthroquinones	Naphthoquinones
Antileukaemic agents	Nucleic acids
Antitumor agents	Nucleotides
Antiviral agents	Oils
Aromas	Opiates
Benzoquinones	Organic acids
Carbohydrates (including polysaccharides)	Peptides
Cardiac glycosides	Perfumes
Chalcones	Phenols
Dianthrenes	Pigments
Enzymes	Plant growth regulators
Enzyme inhibitors	Proteins
Flavanoids, flavones	Steroids and derivatives
Flavours (including sweeteners)	Sugars
Furanocoumarines	Tannins
Hormones	Terpenes and terpenoids
Insecticides	Vitamins

Once cultures have been initiated, a number of possible approaches then exist for the yield inhancement of the products (Fig.10.2). Yield of secondary compounds in cell cultures can be increased by taking the explant from parent plant which accumulates high level of a particular secondary compound. Variant cells in these cultures may be selected which have the capacity to produce and accumulate secondary compounds.

According to Mantell and Smith (1983) the accumulation of secondary metabolites at any one stage of culture is the result of a dynamic balance between biosynthetic, biotransformational and biodegradative process.

Table 10.3. Examples of increase in natural - product yield compared with parent plant material (after Fowler, 1983)

Product	Species	Cell culture Yield (% dry wt)	Whole (field grown) plant yield (% dry wt)	Ratio cell culture/ whole plant
Ajmalicine	*Catharanthus roseus*	2.2	0.3	7
Anthroquinones	*Morinda citrifolia*	18.0	2.2	8
Diosgenin	*Dioscorea deltoidea*	3.5	2.0	2
Glutathione	*Nicotiana tabacum*	1.0	0.1	10
Nicotine	*Nicotiana tabacum*	5.0	2.1	2
Rosmarinic acid	*Coleus blumei*	15.0	3.0	5
Terpentine	*Catharanthus roseus*	1.8	0.5	3

Advantages of Tissue Cultures in Production of Useful Compounds

There are a number of advantages in using plant cell culture for large scale production of industrially important compounds. These are as follows:

1. Plant cell cultures are independent from environmental factors,
2. Production levels may be geared more accurately according to the market demand,

3. By using characterized cell lines, a more consistent product quality and yield can be maintained,
4. New routes of synthesis can be recovered from mutant cell-lines which may lead to the development of novel products, and
5. Culture of cells will reduce the pressure on already over-exploited medicinal and other economically important plants.

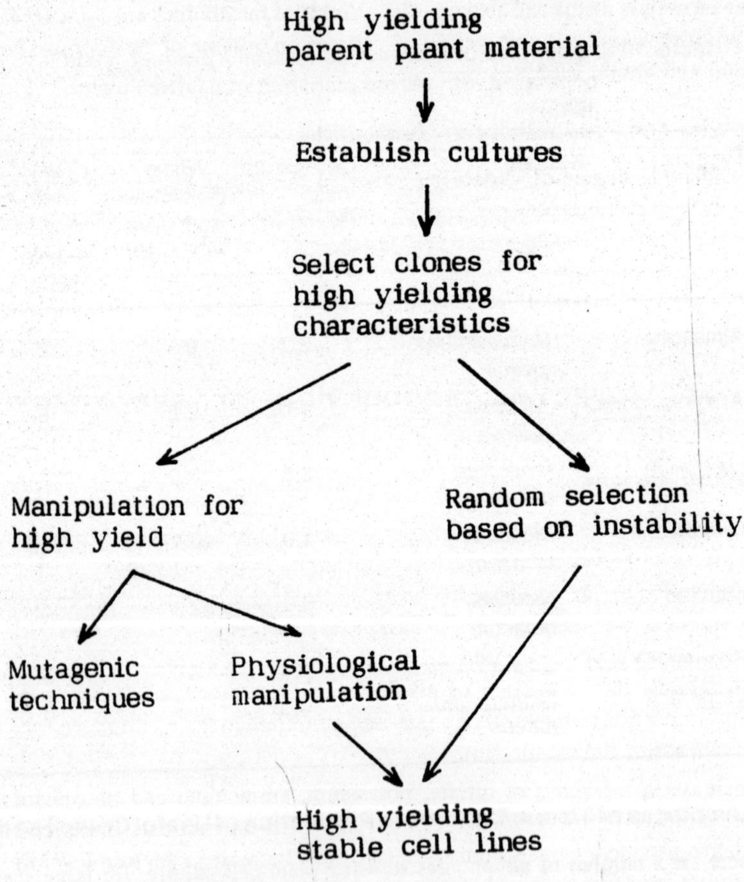

Fig.10.2. Selection strategy for high yielding cell lines.

Culture Techniques

Usually three main types of *in vitro* cultures are employed in the study of secondary plant metabolites: (a) organ cultures, (b) callus cultures (usually grown on solid media) and (c) liquid cell suspension cultures. In the latter, metabolic studies are usually carried after 5-10 generations in order to eliminate the chances of carryover of secondary compounds from parent tissue. Some metabolites, especially alkaloids excrete from the cells in the media, therefore, both qualitative and quantitative analyses of secondary metabolites is done on culture media and also on cells. Yields of metabolites are compared on a product weight per unit weight of cells or on volume of medium basis (Mantell and Smith, 1983).

Genotype and Productivity of Metabolites

A considerable degree of variability exists in the production of secondary metabolites in cultures obtained from different mother plants or even from a single mother plant. This happens due to the genotypic variation in mother plant e.g. in isogenic lines of tobacco, high nicotine yielding mother plants give high yielding cultures whereas low yielding mother plants give low yielding cultures.

FACTORS AFFECTING PRODUCTION OF SECONDARY METABOLITES

1. Culture Medium

The medium which limits rapid cell division and early cessation of exponential growth is best for production of secondary metabolites. Growth regulators play an important role in determining the potential productivity of a given culture. For stimulating alkaloid synthesis in suspension culture of *Papaver bracteatum*, IAA has been found to be better than other auxins. At high concentrations, kinetin inhibits the production of alkaloids in *Datura tabula*. In *Solanum aviculare* reduction in the level of auxin and cytokinin in the medium, results in the increase of steroid spectrum.

Increase in the levels of nitrate, potassium, ammonium and phosphate support rapid cell growth but decrease of any of these nutrients limits cell growth and production of secondary metabolites. Increase in sucrose level in the medium increases the yield of secondary metabolites.

2. Light

Light stimulates biosynthesis of secondary metabolites in cultures. For example, 'cool white' light stimulates biosynthesis of diosgenin in tuber derived callus and cell suspensions of *Dioscorea,* solasodine and solamargine in *Catharanthus roseus* cell cultures.

3. Temperature

Temperature greatly affects secondary metabolite production in cultures. In *Peganum* optimal growth of the callus occurred at 30 ^{0}C but maximum alkaloid production was attained at 25 ^{0}C. The production of alkaloid decreases at higher temperatures.

4. Rotation Speed of Shaker

In tobacco, increase in the rotation speed of shaker (150 r min^{-1})induces nicotine production but normal agitation (110 r min^{-1})results in slight inhibition of nicotine synthesis.

5. pH of Medium

pH of the medium controls the biosynthesis of secondary metabolites. In *Ipomoea* when cells are cultured at pH 6.3, the production of tryptophol becomes double. If pH of the medium drops to 4.8, synthesis of tryptophol is completely inhibited.

11

PROTOPLAST CULTURE AND SOMATIC HYBRIDIZATION

Protoplasts are spherical naked plant cells obtained by the removal of cell wall. Torrey and Landgren (1977) defined protoplasts of higher plants as "cells with their walls stripped off and removed from the proximity of their neighbouring cell." The protoplasts are an excellent tool for the synthesis of novel combinations of genes and are essential part of the overall processes required for the genetic engineering of plants.

ISOLATION OF PROTOPLASTS

Protoplasts can be isolated either by mechanical or enzymatic method.

Mechanical Method

The mechanical method involves isolation of protoplasts by cutting plasmolysed tissues with a sharp razor. The protoplasts are released from the cut ends of the cells. This method has certain limitations viz., (i) only a small number of protoplasts can be isolated, (ii) the method is tedious and time consuming and (iii) it is not useful for isolating protoplasts from mature and meristematic tissues. Advantage of this method lies in the fact that it eliminates the unknown effects of enzymes on protoplasts.

Enzymatic Method

Cocking (1960) for the first time demonstrated the large scale isolation of protoplasts from higher plants. By using this method, protoplasts can be

isolated from any part of the plant body but it is easier to isolate protoplasts from mesophyll of leaf and pollen mother cells or tetrads. Isolation of protoplasts through this method has certain advantages: (i) large quantity of protoplasts can be isolated, (ii) osmotic shrinkage is very little and (iii) the cells are not injured as observed in mechanical method.

Sometimes, the mechanical and enzymatic techniques are combined to isolate protoplasts. In this process, the cells are first separated mechanically and then enzymatically transformed into protoplasts.

Isolation Techniques

The basic technique involved in the isolation of protoplast from mesophyll cells (Fig. 11.2) consists of following steps:

1. The leaves obtained from 7-8 week old plants are surface sterilized. Sterilization is done either by immersing the leaves in 70% ethyl alcohol for 30-60 seconds followed by rinsing in 0.4-0.5% sodium hypochlorite solution or 10% solution of calcium hypochlorite for 15-20 minutes.
2. The leaves are repeatedly washed with sterile distilled water.
3. The lower epidermis of leaves is carefully peeled with the help of forcep and peeled areas are cut into small pieces (4 cm^2).
4. The peeled leaf pieces are placed in enzyme mixture containing 0.5% macerozyme, 0.8 M D-mannitol and 0.3% of potassium dextran sulphate at pH 5.4 and gently shaken in a water bath at 25°C. Fresh enzyme mixture is replaced at 30 minute intervals till mesophyll cells are completely isolated.
5. The peeled leaf pieces are placed in wall degrading enzyme mixture containing 0.5% pectinase enzyme and 2% cellulase in sorbitol or mannitol at pH 5.4 and incubated for 12 hours at 25°C.
6. Segments are gently teased to liberate the protoplasts and filtered to remove debris. The filtrate is centrifuged at 100 g for 1 min. The protoplasts settle at the bottom of the tube. The supernatant enzyme medium and leaf debris are removed. The protoplasts are washed with 0.7M sorbitol and later replaced with 20% sucrose solution. This solution is centrifuged at 200 g for 1 min. The protoplasts are removed with a pipette and placed in a suitable culture medium (Fig. 11.1A)

CULTURE OF PROTOPLASTS

Protoplasts may be cultured either by liquid drop method or plating method. In the liquid drop method droplets of protoplasts suspension are placed in liquid

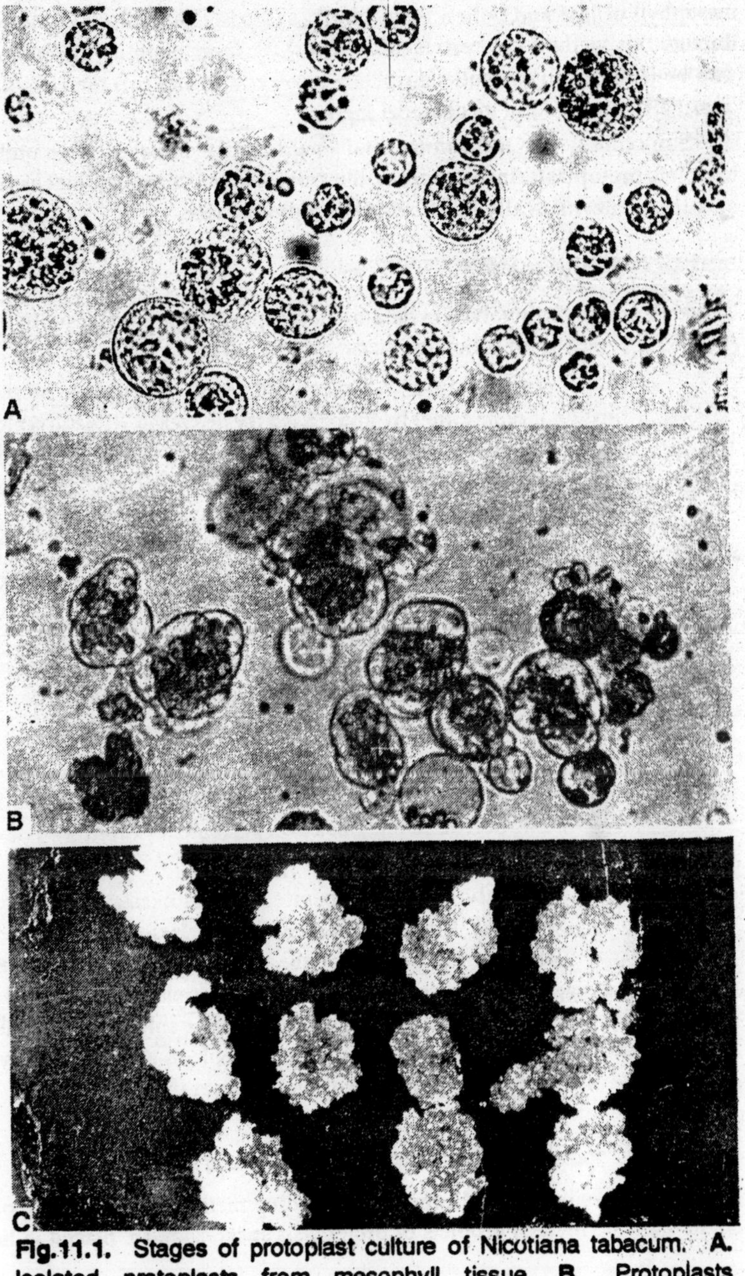

Fig.11.1. Stages of protoplast culture of Nicotiana tabacum. A. Isolated protoplasts from mesophyll tissue. B. Protoplasts showing cell divisions forming multicellular clumps. C. Calli formed from dividing protoplasts at a very initial stage of regeneration on a differentiating medium.

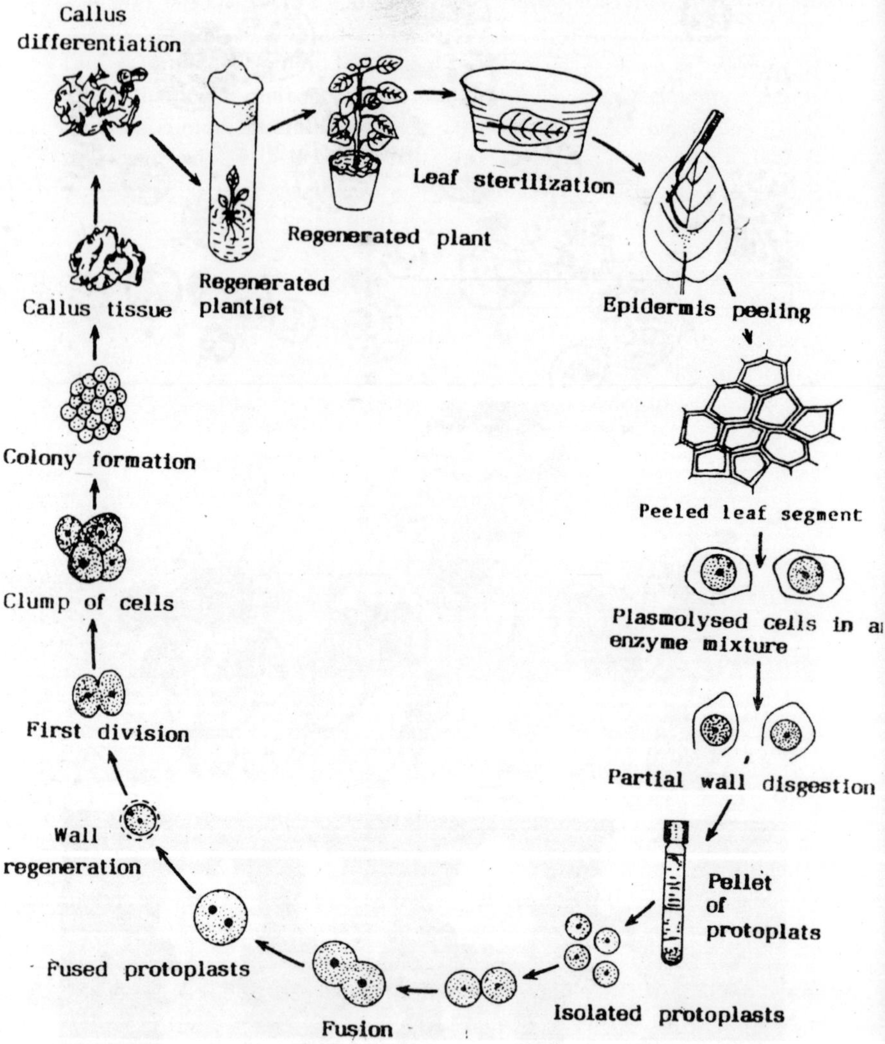

Fig.11.2. Isolation, culture and fusion of leaf cell protoplasts.

medium in micro-chamber. Protoplasts may be transferred on a semisolid medium after a few divisions. Murashige and Skoog's or Nagata and Takebe's media with some modifications are commonly used for culturing protoplasts. Plating method is most commonly used method of protoplast culture. A 2 ml solution of protoplast suspended in liquid culture medium is poured into a petridish and mixed with equal volume of same culture medium containing 1.2% agar. The temperature of medium is maintained at 40°C. The petridishes are sealed with paraffin to avoid dessication. The dishes are incubated in an inverted position at 25-28°C at a continuous light intensity of 2300 lux.

CELL WALL FORMATION AND DIVISION

Regeneration of wall begins after a few hours of culturing the protoplast. The completion of cell wall, however, takes 2-several days. A newly formed cell wall is composed of loosely arranged microfibrils which later on organise to form a typical cell wall (Fig. 11.1B)

Protoplasts increase in size and numerous cytoplasmic strands develop during and before the early stages of cell wall regeneration. There is considerable increase in the number of cell organelles, respiration and synthesis of RNA, protein and polysaccharides. Majority of the cell organelles aggregate around the nucleus.

The synthesis of microfibrils occurs on the surface of plasma membrane. Robenek and Peveling (1977) reported the involvement of endoplasmic reticulum in the synthesis of wall material. The protoplasts divide within 2-7 days of culturing and multicellular colonies are formed after 2-3 weeks.

REGENERATION AND ORGANOGENESIS

Although it is easy to isolate protoplasts but it is difficult to induce them to develop a new cell wall, divide and regenerate a new plant. Takebe *et al.* (1971) for the first time reported regeneration of plants from isolated protoplasts in *Nicotiana tabacum.* Since then plant regeneration has been achieved from cultured protoplasts in members of families Solanaceae, Rutaceae, Apiaceae, Brassicaceae, Euphorbiaceae. (Table 11.1)

Plants from the protoplasts may be regenerated either through embryoid formation or through the production of calli (Fig. 11.1C). In carrot (Daucus carota) protoplasts when cultured on the medium containing 2,4-D produce cell aggregates and give rise to embryoids. These embryoids developed into small plantlets which upon transfer to a 2,4-D free medium, developed into carrot plants.

Protoplasts obtained from *Asparagus officinalis* cladodes when cultured on a medium containing NAA, zeatin and L-glutamine gave rise to small colonies. Root differentiation started when protoplast colonies were transferred

Table 11.1. Plant species in which complete plants have been regenerated from the protoplasts

1. **Apiaceae**	
Apium graveolens	(celery)
Daucus carota	(carrot)
2. **Asteraceae**	
Helianthus annuus	(sunflower)
Cichorium endiva	(endive)
3. **Brassicaceae**	
Brassica rapa	(turnip)
B. napus	(rapeseed)
B. oleracea	(cabbage)
4. **Fabaceae**	
Medicago sativa	(alfalfa)
Lotus corniculatus	(birdsfoot trefoil)
Trifolium pratense	(red clover)
5. **Poaceae**	
Triticum aestivum	(wheat)
Zea mays	(maize)
Saccharum officinarum	(sugarcane)
Pennesitum americanum	(pearl millet)
Lolium multiflorum	(Italian rye grass)
6. **Solanaceae**	
Lycopersicon esculentum	(tomato)
Nicotiana tabacum	(tobacco)
Solanum melongena	(egg plant)
S. tuberosum	(potato)
Petunia spp.	(petunia)

to a medium containing reduced level of glutamine. Shoot formation began when zeatin was substituted for BAP and adenine was added in the medium. Variations in the level of hormones resulted in the regeneration of embryoids from protoplasts. When these embryoids were transferred on semi-liquid medium containing IAA and zeatin, they developed into whole plants.

In *Atropa belladona* isolated protoplasts from cell suspension cultures, when cultured developed cell wall around them and first divided within 2-3 days resulting in the formation of small colonies.These colonies further divided to form callus. The callus when grown on NAA-free liquid medium produced somatic embryos. These embryos developed into plantlets on an auxin free medium.

The regeneration of entire plant from protoplasts obtained from mesophyll cells in *Nicotiana tabacum* was first demonstrated by Nagata and Takebe (1971) and later on by Vasil and Vasil (1974). Isolated protoplasts produced cell colonies which subsequently developed into callus. Upon transfer to MS medium supplemented with IAA and KN, callus developed plantlets. Like tobacco, in *Petunia* also plants have been regenerated from protoplast induced calli.

SOMATIC HYBRIDIZATION

The technique of hybrid production through the fusion of protoplasts from different genetic backgrounds is known as *somatic hybridization* or *parasexual hybridization.* The technique of somatic hybridization is of special significance for the improvement of the vegetatively propagated crops such as banana, cassava, potato, sweet potato and sugarcane. Production of somatic hybrids involves a number of steps (Fig. 11.3).

Protoplast Fusion

The fusion of protoplasts may be of two types: (a) spontaneous fusion and (b) induced fusion. In the former, protoplasts may fuse spontaneously during isolation whereas in latter protoplasts are induced to fuse. In somatic hybridization, spontaneous fusion is of no significance and only induced fusion has received popularity. To induce protoplast fusion suitable chemical agents, called as fusogen are used. These include sodium nitrate, polyvinyl alcohol, dextran or polyethylene glycol (PEG). Among all the fusogens, PEG has received wide acceptance because of high frequency of heterokaryon formation and low cytotoxicity. The PEG-induced fusion is non-specific and has been successfully used to fuse protoplasts from wide variety of plant species.

Another method of protoplast fusion is by using electric fields. When protoplasts are introduced into a high frequency alternating electric field, they act as dipoles and are attracted to each other. Once the protoplasts come in contact, a high frequency of fusions may be achieved by disrupting the protoplast membranes with a short direct current pulse which unlike chemical fusogens, have relatively little effect on protoplast viability. This method has now been demonstrated to provide high frequency of fusion.

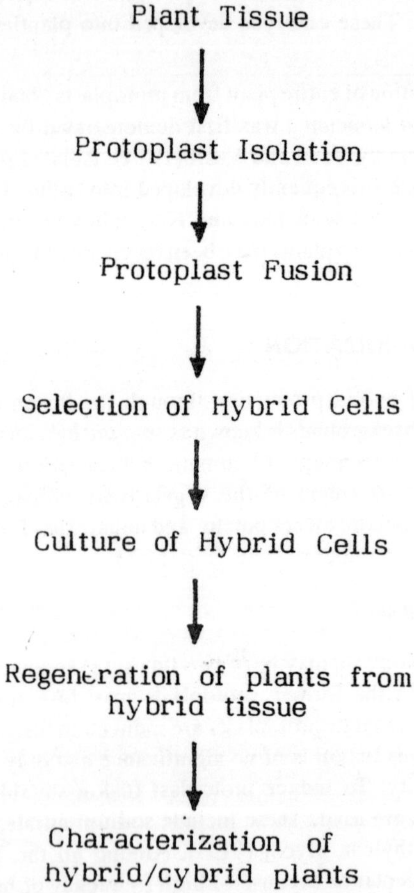

Fig.11.3. Sequential steps involved in somatic hybridization.

Fusion of protoplasts includes three distinct stages (Fig.11.4). During agglutination the plasma membrane of protoplasts come in close contact with each other followed by membrane fusion at localized places. This results in the formation of cytoplasmic channels. The fused protoplasts begin to become round due to expansion of cytoplasmic bridges thus forming spherical hetero- or homokaryons (Fig. 11.5). Fusion between the isolated protoplasts of same species gives homokaryons whereas fusion between different species results in heterokaryons.

Carlson *et al.* (1972) for the first time fused the leaf protoplasts of *Nicotiana glauca* and *N. langsdorffii* and successfully regenerated interspecific hybrid plant. Since then several interspecific and intergeneric hybrid plants have been raised through protoplast fusion (Table 11.2).

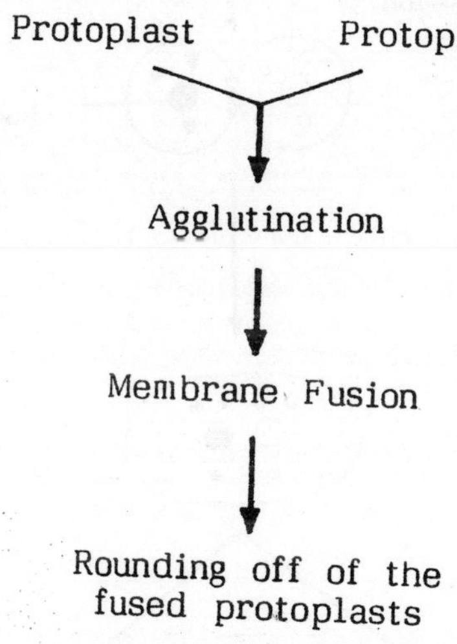

Fig.11.4. Schematic presentation of protoplast fusion.

SELECTION PROCEDURE

In somatic hybridization, the most important aspect is to use certain selective markers for the identification of heteroplasmic fusion products. In a cross between tobacco and carrot, green protoplasts of the former and red anthocyanin of latter were used as markers. Similarly, protoplast fusions between rapeseed and soyabean plastids have been used as selective markers. Somatic hybrid plants can be characterized on the basis of morphology, chromosome number, isozyme pattern, chloroplast DNA and ribosomal RNA analysis.

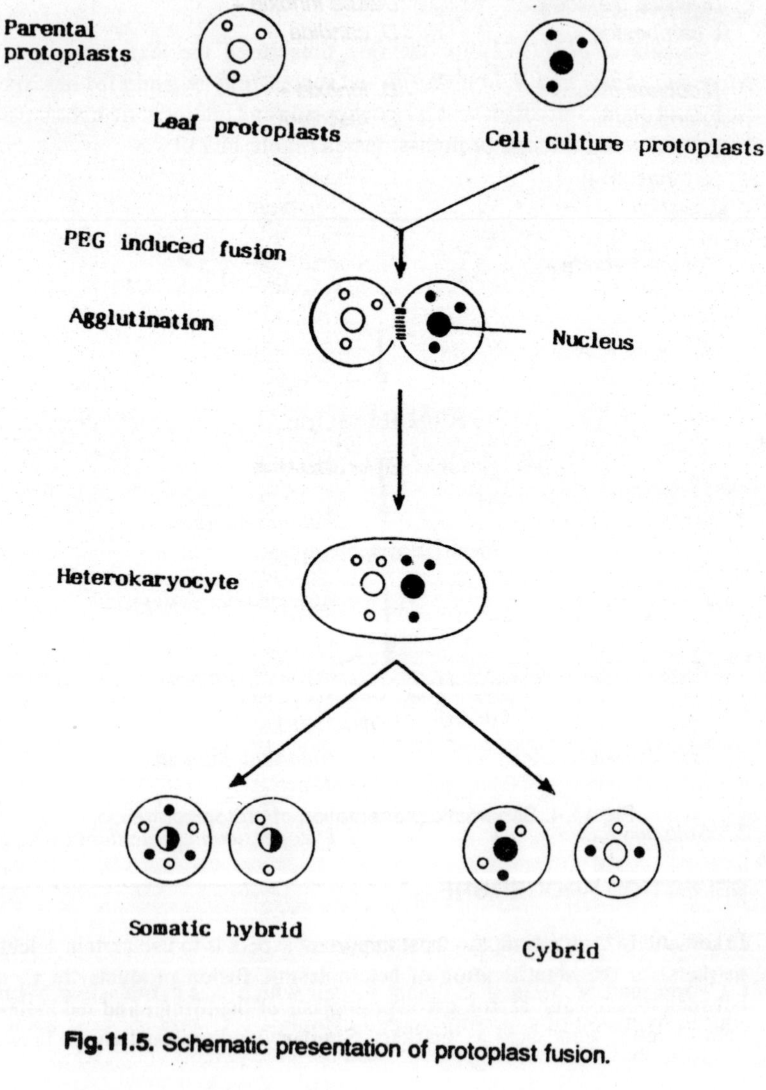

Fig.11.5. Schematic presentation of protoplast fusion.

Table 11.2. Examples of somatic hybrids raised through protoplast fusion

Sexually compatible	Sexually incompatible
Interspecific Hybridization	
1. *Brassica oleracea* + *B. campestris*	*Datura innoxia* + *D. candida*
2. *Nicotiana glauca* + *N. langsdorfii*	*D. innoxia* + *D. discolour*
3. *N. tabacum* + *N. rustica*	*D. innoxia* + *D. stramonium*
4. *Petunia hybrida* + *P. parodii*	*Nicotiana sylvestris* + *N. knightiana*
5. *Solanum tuberosum* + *S. nigrum*	*Petunia parodii* + *P. parviflora*
Intergeneric Hybridization	
1. *Atropa belladona* +	*Petunia hybrida*
2. *Daucus carota* +	*Aegopodium podagraria*
3. *Datura innoxia* +	*Atropa belladona*
4. *Hyoscyamus muticus* +	*Nicotiana tabacum*
5. *Solanum tuberosum* +	*Lycopersicon esculentum*

CYTOPLASMIC HYBRIDS (CYBRIDS)

Cybrid may be defined as "plant or cell which is a cytoplasmic hybrid produced by fusion of a protoplast and cytoplast". Cybrids possess nuclear genes from only one and cytoplasm from both the parents (Fig.11.5).

In cultures, cybrids may arise by one of the following means: (a) removal of one of the nuclei after the formation of heterokaryon, (b) fusion between a normal protoplast with an enucleate protoplast, (c) fusion between normal protoplast with a protoplast having a non-viable nucleus, and (d) selective loss of genome of one of the nuclei.

Power *et al.* (1975) fused mesophyll protoplasts of *Petunia* with cultured cell protoplasts of the crown gall of *Parthenocissus* and isolated a line which contained the chromosomes of only *Parthenocissus* but showed some of the properties associated with *Petunia* for some time. It has been observed that usually a selective elimination of the plastids of one of the partners takes place. Using certain selection pressures, Zelcer *et al.* (1978) transferred cytoplasmic male sterility from *Nicotiana tabacum* to *N. sylvestris*. Similarly, Medgyesy *et al.* (1980) transferred streptomycin resistant character of *Nicotiana tabacum* to *N. sylvestris* by fusing the non-dividing protoplasts of former with normal protoplasts of latter. The importance of protoplast fusions for transfer of cytoplasmic inherited genes of both parents is of special significance in crop breeding.

Maliga *et al.* (1982) fused enucleated protoplasts with nucleated protoplasts in *Nicotiana*. Protoplasts of *Nicotiana tabacum* (SRI) carrying a maternally inherited streptomycin - resistant mutation were enucleated. The resulting cytoplasts were then fused with streptomycin sensitive protoplasts of *Nicotiana plumbaginifolia*. The SRI cytoplasts having no nuclei could not form callus. All resistant clones recovered after fusion were supposed to be derived from interspecific cytoplast protoplast fusion. Plants regenerated from eight of the resistant clones were resistant to streptomycin and inherited the resistance maternally.

Cybrids could be exploited for transferring male sterility from one species to another, especially where the sexual crosses are incompatible.

APPLICATIONS OF SOMATIC HYBRIDIZATION

1. Novel hybrid plants can be produced in sexually incompatible species.
2. Heterozygous lines may be developed within a single species which are usually propagated by means of vegetative methods.
3. Disease resistance can be induced by incorporating selective genome from the disease resistant to susceptible crop plants.
4. Photosynthetic efficiency of plants can be enhanced through transplantation of foreign chloroplasts in plants having less efficient photosynthetic systems.

5. Cytoplasmic male sterile lines can be developed through fusion of an 'elite' cultivar with the x-ray-irradiated protoplasts of CMS parent.
6. Limited parts of genomes can be transferred from one species to another by the formation of heterokaryons in which unidirectional sorting of cytoplasmic elements takes place.

PROTOPLAST CULTURE AND GENE TRANSFERS

Rapid advances made in the field of molecular biology and genetic engineering have provided a rare opportunity for transfer and interchange of genes of plants as well as animals. The genetic transformation achieved in higher plants through the incorporation of foreign DNA has greatly helped in creation of novel genotypes. Recombinant DNA techniques provide an unlimited gene pool for crop improvement. Introduction of foreign genetic material involves two different approaches: (1) use of vectors and (2) transformation through direct DNA introduction. The gene transfer technology would shorten the time taken for breeding but will not replace it.

DIRECT DNA UPTAKE BY PROTOPLASTS

Protoplasts, being wallless in nature, are ideal system for the introduction of foreign genetic material. Uptake of exogenous DNA by protoplasts has been reported in petunia, soybean, carrot and other plant species. The transformation of higher plants by introducing DNA into their cells involves a highly complex process. This includes uptake, integration, replication, expression and transmission of the introduced DNA.

Protoplasts are used as recipients of genetic transfer, therefore, regeneration of plants from transformed protoplasts becomes important. In tobacco, transformation and Mendelian segregation of the transferred trait has been reported by Paszkowski *et al.* (1984). Transferred trait is inherited as dominat gene. Direct DNA uptake by protoplasts has been successfully demonstrated in *Triticum monococum* and *Lolium multiflorum.*

There are four modes of DNA delivery into the protoplasts. These are: microinjection, liposome encapsulation, electroporesis and shot gun method.

Microinjection: This method involves the transfer of DNA solutions under pressure into plant protoplasts by means of micropipettes (Fig. 11.6). Microinjection method has certain advantages over other modes of DNA uptake: (a) high frequency of transformation and (b) minimum somaclonal variation. This method is being utilized for the transformation of gametes and the transfer of chromosomes and organelles between plant species.

Intranuclear injection of tobacco or alfalfa protoplasts can result in a transformation frequency of about 15-25%.

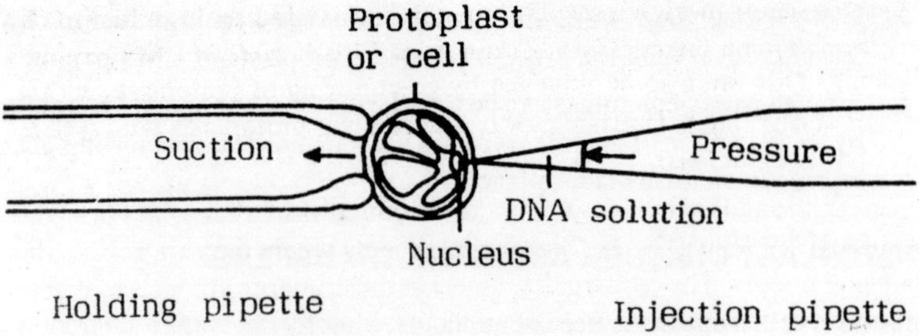

Fig.11.6. Microinjection of protoplasts or cells.

Liposome encapsulation: Liposomes are artificial lipid vesicles of uniform size (0.4 μm diameter or more) which can be prepared by sonicating a mixture of phosphatydyl serine and an aqueous buffer. DNA can be encapsulated in this unilamellar vesicle and may be delivered to the protoplasts. Tobacco protoplasts have been successfully transformed by using liposomes.

Electroporesis: In this method, high voltage is used at the time of protoplast fusion for a very short period. During electroporation, reversible breakdown of the cell membrane, increase in conductivity and permeability take place. The transient permeability promotes the exchange of intra- and extracellular compounds by diffusion. Electroporation has been successfully used for introducing genes into mammalian cells. It is now being used for both the transient and stable transformation in cereals.

Shot-gun method of DNA introduction : Klein *et al.* (1987) developed a novel method of introducing DNA into cells of cultured explants capable of regenerating to give somatic plants. In this method extremely fine metallic pellets (tungsten micro-projectiles; spherical particles, 4-μm diameter) coated with DNA, are shot in the plant tissue by means of a 0.22 calibre gun. These pellets help in direct delivery of DNA into the cells. The pellets do not damage the plant tissue. Some recent experiments done on the onion plants have shown that when bacterial and viral genes were fired, such onion plants were found to start producing the foreign proteins.

AGROBACTERIUM MEDIATED GENE TRANSFER

Agrobacterium tumefaciens is a soil bacterium which causes crown gall tumors in dicotyledons. These tumors are formed due to insertion of Ti plasmids into the nuclear genome of the infected plant (Fig. 11.7). The Ti plasmid contains T- DNA within it. The T- DNA causes hormonal disturbances in the transformed plant cells. Most notable of these are increased level of growth hormones, auxins and cytokinins. The capacity of *A. tumefaciens* to cause tumor formation can be removed by disarming the strain through deletion of genes in the T- DNA without loss of DNA transfer and integration functions.

Gene transfer in higher plants through Ti plasmid can be achieved by tumor formation on intact plants or plant parts with *A. tumefaciens* carrying a Ti plasmid. This can also be achieved by co-culturing of protoplasts with *A. tumefaciens* carrying a Ti plasmid or fusion of plant protoplasts with sphero-plasts of *A. tumefaciens*.

The Ti plasmids act as vectors for the introduction of alien genes in a wide range of higher plants (Fig. 11.7) except monocots where they are not capable of inducing tumors. The crown gall tumors synthesize large amounts of certain metabolites, called opines. These metabolites are not formed in normal tissues. The genes which are responsible for the synthesis of opines are borne on the Ti plasmid and can integrate into the genome of host cells.

The transformation system involving *Agrobacterium* includes replacement of the natural tumor inducing genes by bacterial genes. These genes are

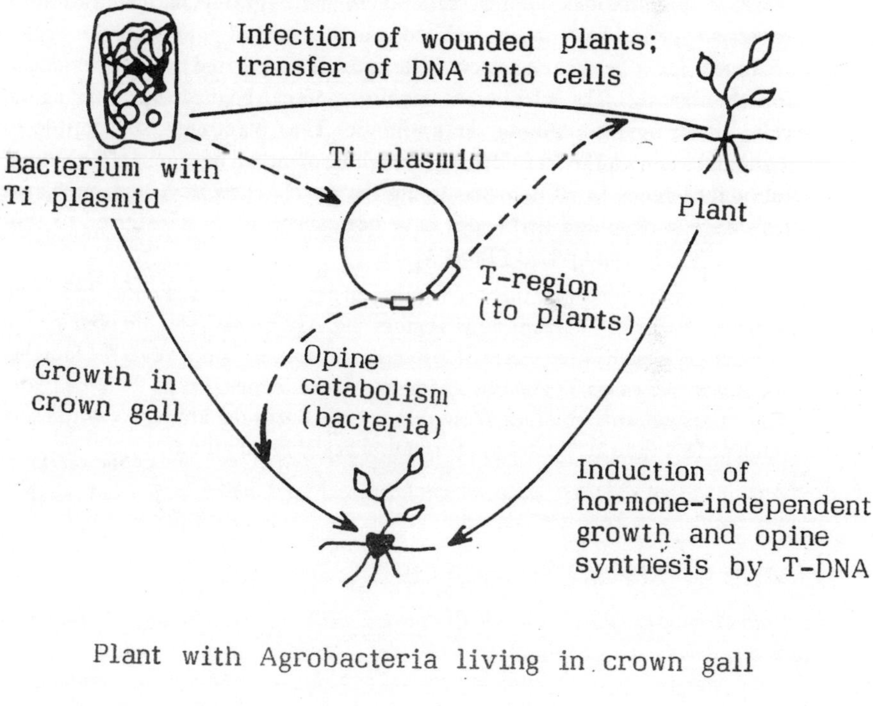

Fig.11.7. Genetic colonization of plants by *Agrobacterium tumefaciens*.

Table 11.3. Some economically important plants transformed via *Agrobacterium*

Sugar beet	(*Beta vulgaris*)
Turnip	(*Brassica rapa*)
Carrot	(*Daucus carota*)
Sunflower	(*Helianthus annuus*)
Walnut	(*Juglans regia*)
Cowpea	(*Vigna unguiculata*)
Potato	(*Solanum tuberosum*)
Tobacco	(*Nicotiana tabacum*)

resistant to antibiotics such as kanamycin and hygromycin. The antibiotic resistance genes have been isolated and engineered into suitable vector plasmids. These *in vitro* engineered plasmids are transferred by *Agrobacterium* into the plant cell. The selection of transformants can be made by culturing the cells on medium containing the antibiotic. Thus plant cells susceptible to *Agrobacetrium* can be transferred and a whole plant can be regenerated having introduced genes in all its cells. Using the *Agrobacterium* vector, genes for resistance to pests and herbicides have been successfully transferred to economically important plants (Table 11.3).

Successful incorporation of foreign genes in some monocots has been achieved by using Ti plasmids as vectors e.g. *Asparagus*. The *Agrobacterium* infected cells have been observed to synthesise opines. This reveals the fact that the genes carried on Ti plasmids have expressed themselves in the host plant. The callus cultures obtained from tumor cells retain the ability to synthesise opines. This indicates that transformation is stably inherited.

12

RECOMBINANT DNA TECHNOLOGY

The discovery of enzymes that could cut DNA at internal specific sites was an event of great significance because it led to establishing new branch of science called recombinant DNA technology or genetic engineering. The importance of *in vitro* genetic manipulations lies in the fact that it provides an ability to by-pass the multiplicity of mechanisms that restrict gene transfer between unrelated organisms. Recombinant DNA technology includes techniques of cutting DNA into specific fragments using enzymes restriction endonucleases and joining the fragments with the help of other enzyme, ligase. Using this technology, the fragments of foreign DNA can be inserted into a vector.

Plasmids and viruses can incorporate small pieces of host DNA in their own chromosomes and act as vectors for transference of these genes to closely related bacterial species. These vectors bearing chromosomal genes can either become integrated into the chromosome of the recipient cell or exist as independently replicating units in the cytoplasm. Recombinant DNA technology generally involves the following steps:

1. DNA fragments coding for proteins of interest are synthesized chemically or isolated from an organism.
2. These DNA fragments are inserted in a restriction endonuclease cleavage site of the vector that does not inactivate any gene required for the vector's maintenance and selective marker.
3. The recombinant DNA molecules are then introduced into a host to replicate using the replication origin of the vector.
4. Recipient host cells that have acquired the recombinant DNA are selected. Selection pressure is applied to enrich bacteria with a selectable marker.

5. Desired clones are then characterized to ensure that they maitain true copies of the DNA segment that was originally cloned.

CLONING VEHICLES

PLASMIDS

Plasmids are extrachromosomal, self replicating and stably- inherited double-stranded circular DNA molecules resident in many bacterial or even other cells. A variety of plasmids are known which operate as very efficient vectors of genetic information in wide variety of organisms. Plasmids have been utilized in making a large number of copies of a particular gene. Using such cloning plasmids, genomic libraries are established by establishing a genome with suitable endonucleases and cloning the resultant fragments. Information about some of the best known plasmids are detailed below.

Ti plasmids

Ti plasmids are found in tumor cells produced by *Agrobacterium tumefaciens.* The tumor inducing capacity of the bacterium is located in large extrachromosomal plasmids, which are knwon as Ti plasmids (Fig. 11.7). Sterile tumor cells contain a part of the Ti plasmid which is covalently integrated into the plant chromosomes. In bacterium, this part is called T- region whereas in plant cells it is known as T- DNA.

Tumor cells synthesise a number of low molecular weight compounds called opines (Fig. 12.1). These compounds are, however, absent in normal cells. The genes which are responsible for opine synthesis are borne on the Ti plasmid and not on the main chromosome of the *Agrobacterium.* Opines produced are selectively used by the bacteria as sources for carbon, nitrogen and energy.

Ti plasmids are natural gene vectors which contain all the functions essential for the transfer, stable incorporation and expression of genetic information in plants.

Plasmid pBR 322

Bacterial plasmid pBR 322 is one of the best known cloning vehicles. This vector was created artificially by using different parts of certain naturally occurring plasmids. This plasmid vector carries genes for resistance to ampicillin and tetracycline which act as markers for the selection and identification of clones carrying the plasmid (Fig. 12.2). pBR 322 has sites for a variety of restriction endonucleases which are available for insertion of DNA fragments. Taken together its low molecular weight (4.3 kbp; 28 MD; 1.39 μm) it becomes

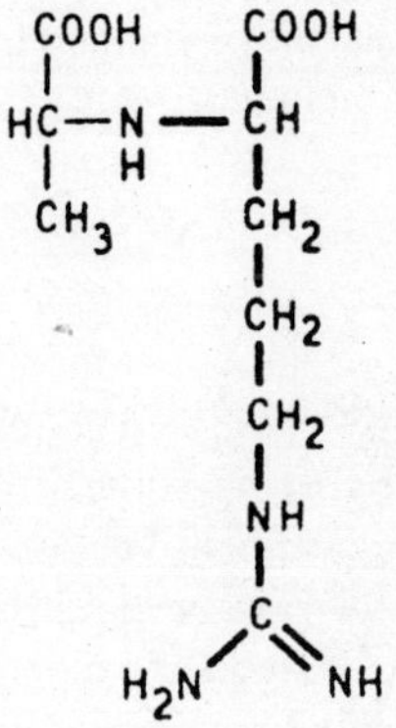

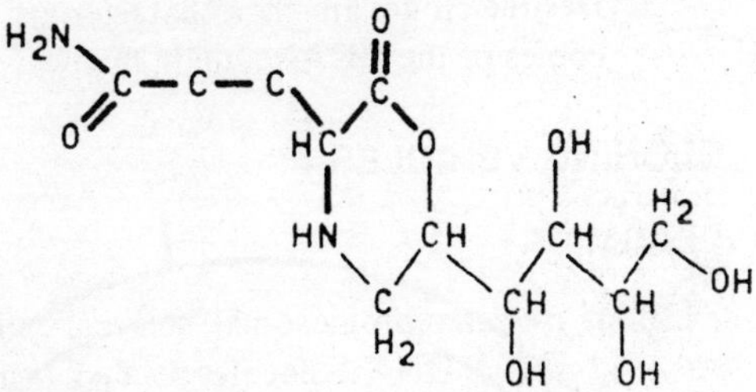

Fig.12.1. Chemical structure of some opines found in crown gall tumors.

evident why this plasmid has been used for gene cloning throughout the world since its creation. The principle of gene cloning with this plasmid is shown in figure 12.3.

Isolation and Purification of Plasmid DNA

Methods of plasmid isolation can vary according to host organism. In *E. coli*, many methods are used for isolating plasmid DNA free from chromosomal DNA. Plasmid-harboring *E. coli* is grown in any rich medium such as Trypticase Soy broth or a minimal salt solution supplemented with casamino acids.

For isolation of plasmid DNA, host cells are treated with lysozyme, EDTA and detergents which weaken the cell wall and plasma membrane. After removal of cellular DNA by centrifugation, plasmid DNA is purified from the cleared lysate. In the alkaline lysis procedures, the DNA is first denatured and upon neutralization, circular interwined plasmid DNA is rapidly renatured. In such cases, the host DNA remains denatured and is removed alongwith cellular materials. From the cleared supernatant, plasmid DNA (alongwith RNA) is precipitated either with ethanol or isopropanol. RNA can be removed by using enzyme RNAse. Plasmid DNA thus obtained can be used for restriction enzyme mapping, subcloning and hybridization with specific probes. Highly purified plasmid DNA can be obtained by isopycnic centrifugation in a CsCl gradient.

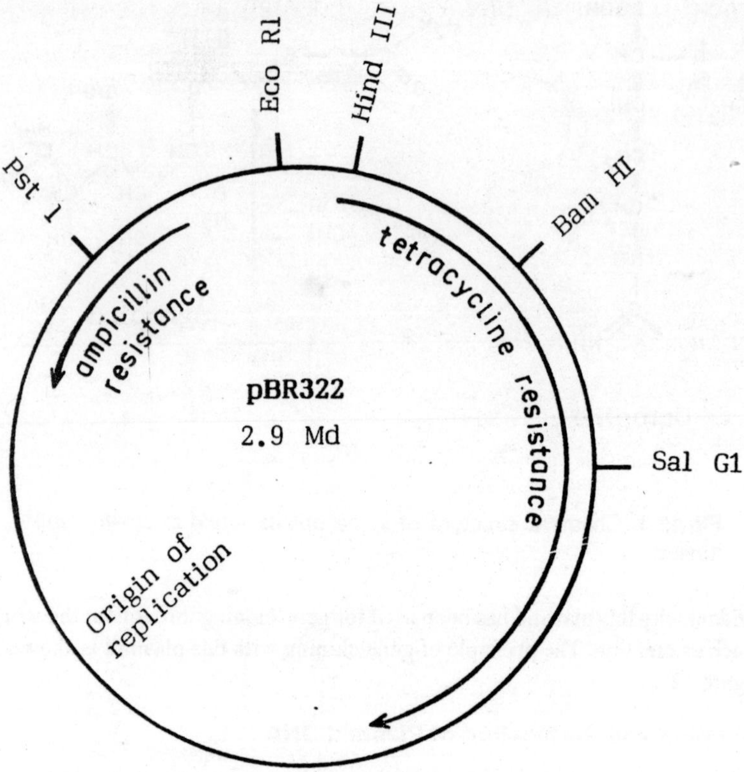

Fig.12.2. Structure of plasmid pBR 322.

BACTERIOPHAGES AND COSMIDS

One of the most extensively studied *Escherichia coli* viruses is the bacteriophage lambda (Fig.12.4). This is a common vector which has a linear double stranded DNA (50 Kbp in size). Isolation of lambda - DNA is more complicated than isolation of plasmid DNA. The yield in this case is also very low. Most genomic libraries are made from this vector because of the large size of the insert that can be accommodated. The 5' ends of the double stranded lambda genome exhibit short single-stranded regions which are 12 nucleotide in size. By the association of these cohesive termini upon injection into the host cell the cos site is formed and the lambda DNA adopts a circular shape. The *in vitro* packaging of the recombined DNA molecules into phage particles markedly

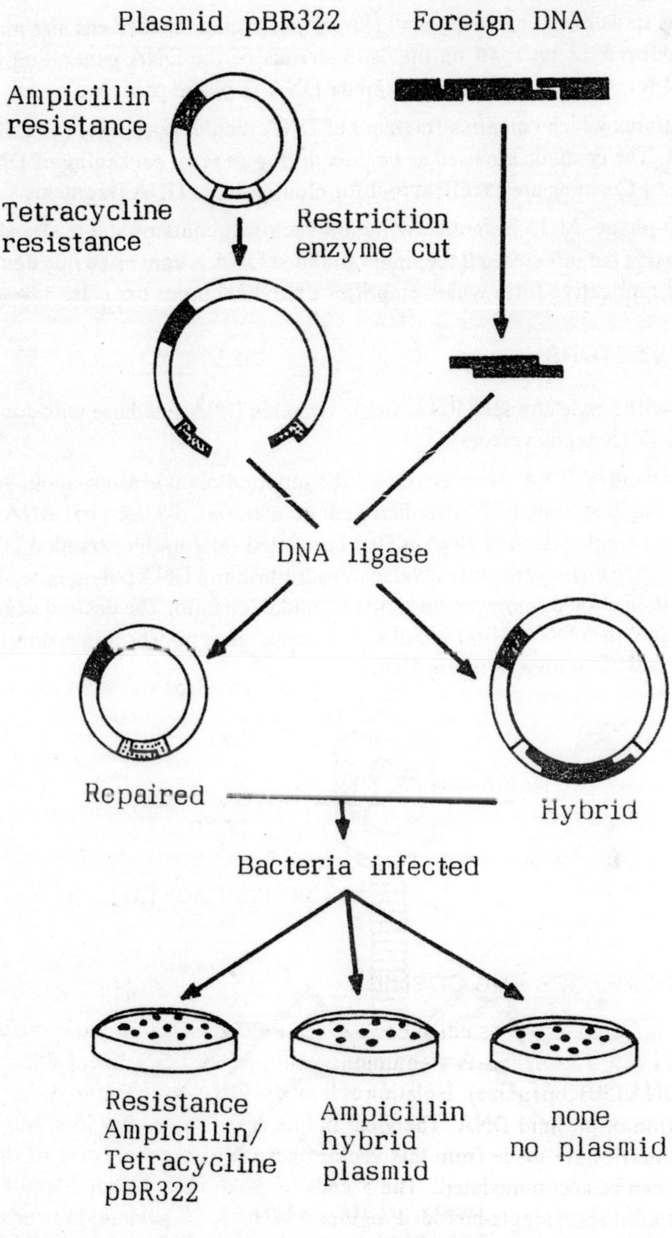

Fig.12.3. Principle of gene cloning using the *E. coli* plasmid pBR 322.

enhances transformation frequency. During packaging, at each **cos site** nicks are introduced 12 bp apart on opposite strands of the DNA generating the cohesive termini that are found in lambda DNA in phage particles.

Plasmids which contain a fragment of DNA including **cos site** are called **cosmids.** The cosmids are used as vectors during *in vitro* packaging of DNA (Fig. 12.5) Cosmids are excellent tool for cloning large DNA fragments.

Coli phage- M 13 is another common vector. It contains single stranded DNA. Inside the infected cell the single stranded DNA is converted into double stranded replicative form which amplifies upto 300 copies per cell.

VIRUS VECTORS

Viruses with single stranded RNA, single stranded DNA and those with double stranded DNA act as vectors.

Single-stranded RNA Viruses: Of all the reported plant viruses, about 90% contain single-stranded RNA as their genetic material. To use viral RNA as vector, the single stranded RNA is first converted into double stranded DNA (copy DNA) by using enzymes reverse transcriptase and DNA polymerase. The DNA is then cloned into a prokaryotic plasmid or cosmid. The desired gene is inserted into the cDNA. Viral vector and 'foreign' gene may be reintroduced in the host cell as shown in figure 12.6.

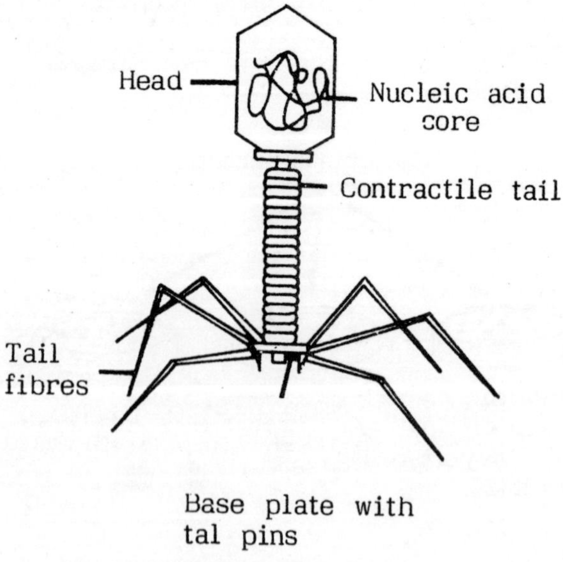

Fig. 12.4. Diagrammatic representation of bacteriophage.

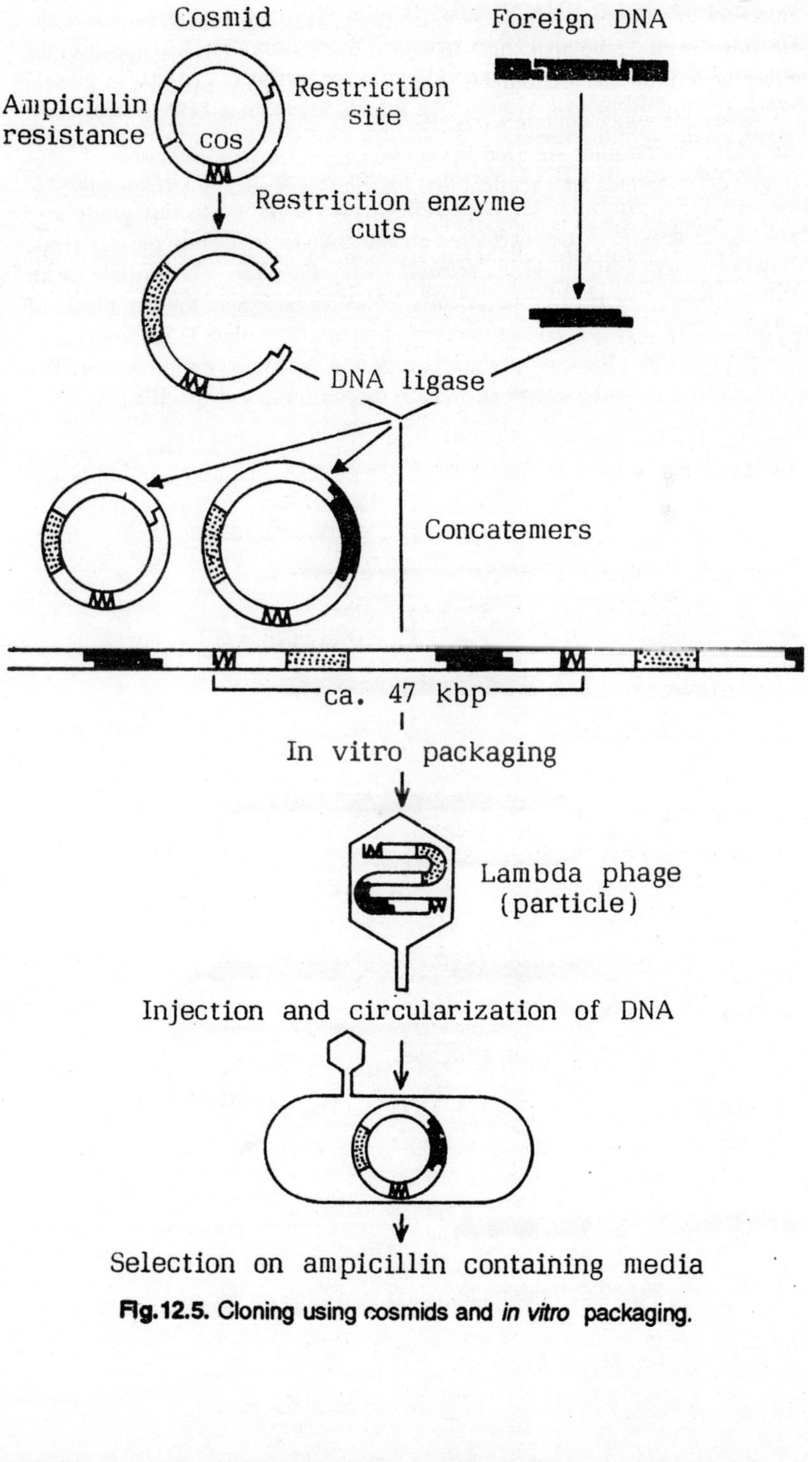

Fig.12.5. Cloning using cosmids and *in vitro* packaging.

Single-stranded DNA Viruses: Gemini viruses are grouped under the category of single-stranded DNA viruses. These viruses are host specific; the host range may be broad (e.g. curley top virus) or narrow (e.g. *Abutilon* mosaic virus). Bean gold mosaic virus which infects legumes is best characterized among all the gemini viruses.

Double-stranded DNA Viruses: Caulimovirus group contains double stranded DNA. Some of the important viruses included in this group are: cauliflower mosaic virus, carnation etched ring virus, Dahlia mosaic virus, Strawberry vein banding virus, *Mirabilis* mosaic virus. The particle of all caulimoviruses occur in proteinaceous inclusion bodies in the cytoplasm of majority of the mesophyll cells of infected plants. To isolate DNA from caulimovirus particles, the coat protein is digested with enzyme proteases. The isolated DNA is double-stranded and has discontinuties at specific sites.

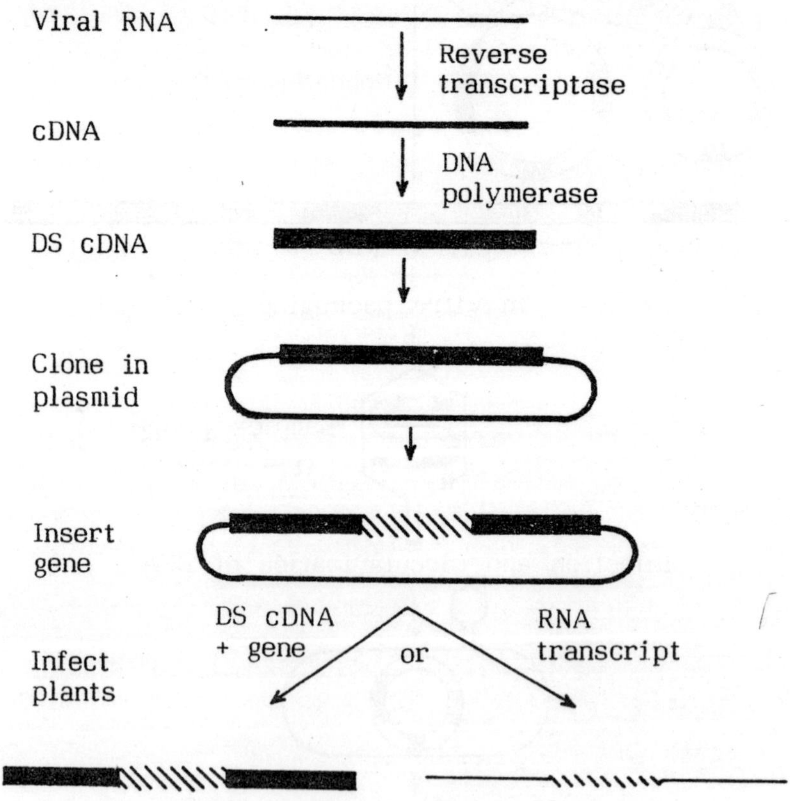

Fig. 12.6. Scheme for using viral RNA as vector.

CLONING STRATEGIES

Methods for the cloning of specific DNA sequences involve insertion of these sequences into vector DNA which is capable of replicating within a host cell. The strategies of gene cloning in bacteria comprises four main stages viz; preparation of the gene, insertion into vector, transformation of host cell and detection of the cloned gene.

1. Preparation of the Gene

Gene cloning in bacteria is achieved by cleaving the DNA with enzyme restriction endonuclease which produces small fragments (approximately 4 kilobase pairs). Each fragment has a 'sticky' complementary single-stranded end (Figs. 12.3, 12.7). Eukaryotic genes contain introns that are not processed in bacteria therefore, DNA for cloning is usually obtained as a reverse transcriptase generated copy DNA (cDNA) of the relevant mRNA. In cases where nucleotide or amino acid sequences are known, synthetic DNA may be produced.

2. Insertion into Vector

The vector is an agent which is used to transfer DNA into a host cell e.g. plasmid, bacteriophage. The vector is cut with the same enzyme (restriction endonuclease) as that used to generate the chromosomal DNA fragments. The chromosomal frgments and linearised vector are incubated with DNA ligase which covalently joins the DNA molecules (Fig. 12.7). Those plasmids which contain an inserted fragment are called recombinant plasmid.

3. Transformation of Host Cell

The ligated plasmid mixture is introduced into the bacterial cell where they takeup DNA through transformation process. Transformation is generally carried out by placing actively growing cells of a bacterium in cold, dilute solution of $CaCl_2$ which enhances the ability of bacterial cells to takeup foreign DNA. In majority of the cases, *E. coli* is the most preferred host because (i) molecular biology of this bacterium is well understood, (ii) calcium chloride treated cells are highly transformable and (iii) *E. coli* transcribes and translates most Gram-positive and Gram-negative genes except some actinomycete genes.

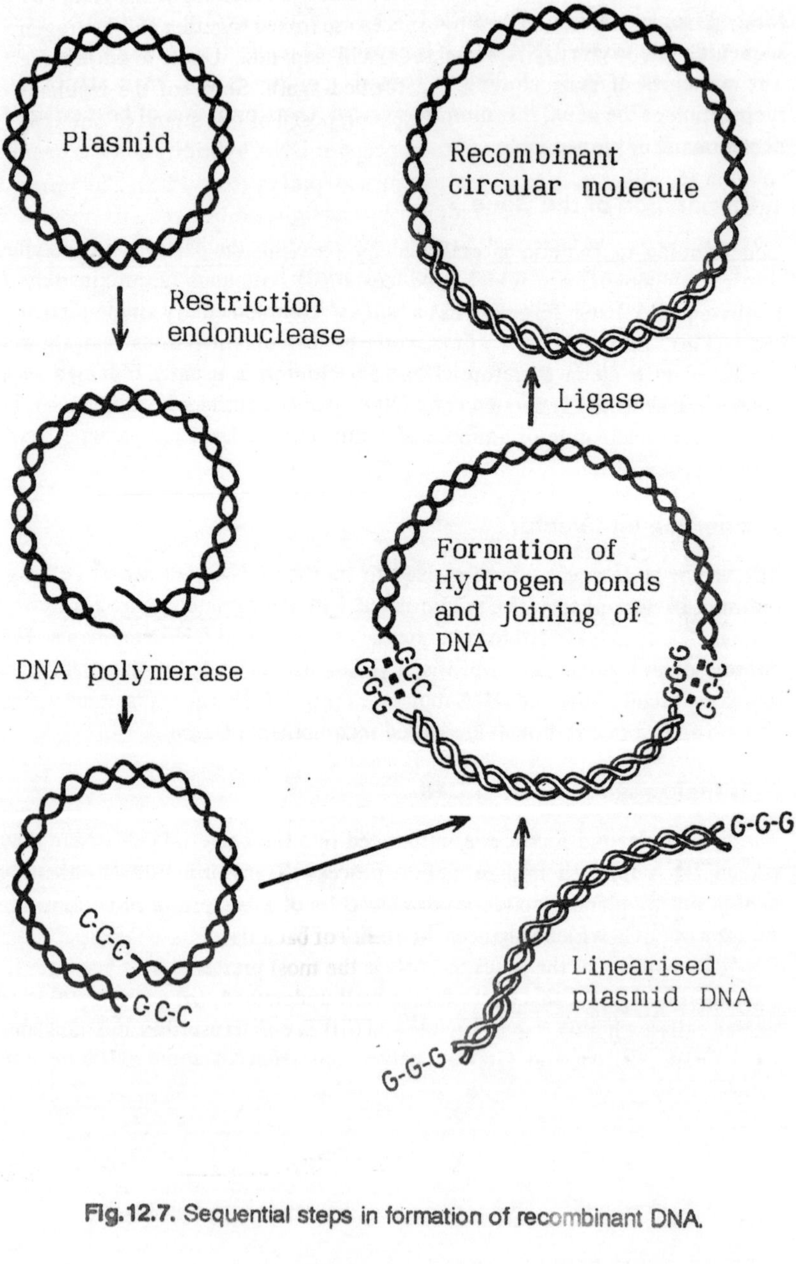

Fig.12.7. Sequential steps in formation of recombinant DNA.

4. Detection of the Cloned Gene

To identify recombinants, one of the important approaches is to use DNA probe. In a DNA molecule, the two complementary strands are held together by hydrogen bonds. If two similar DNA pieces are mixed together and hydrogen bonds broken (by heating) the strands will separate. Upon lowering the temperature, the hydrogen bonds are formed again. Some of the resultant double-stranded DNA will be hybrids i.e. composed of one strand of one type and one strand of the other type. This concept of DNA hybridization has been exploited for utilizing the DNA molecules as probes (Fig. 12.8). The transformed colonies are replica plated to a nitrocellulose filter and are lysed to release the DNA. This DNA is denatured (by raising the temperature) and fixed to the nitrocellulose so as to produce a DNA print corresponding exactly to the position of the colonies on the original plate. The DNA print is then hybridized with the probe which has been previously radioactively labelled. After washing off unhybridized DNA, the position of the radioactive spots on the filter is indicated by autoradiography in order to identify the presence of the required DNA.

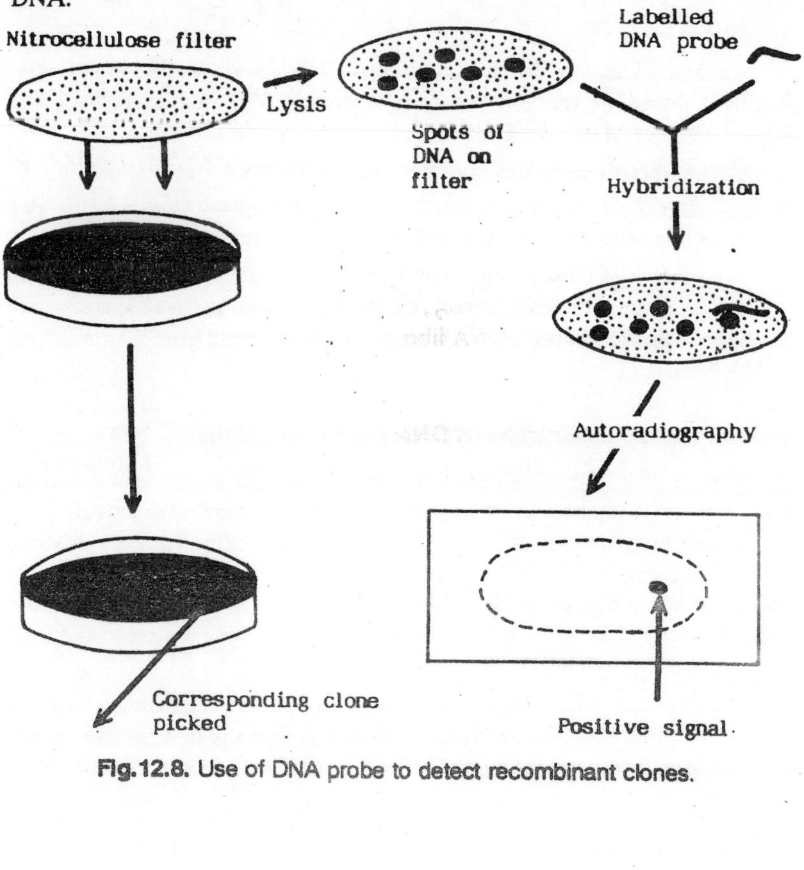

Fig. 12.8. Use of DNA probe to detect recombinant clones.

Genomic and cDNA Libraries

Genomic libraries are also called gene banks. A complete library of genome fragments is a set of independent clones, that, statistically, contains the entire genome among the recombinant DNA molecules. Complete sets of DNAs can be maintained if no segment is disruptive to the vector or host.

Construction of genomic libraries involves a number of steps: (1) the generation of DNA fragments, (2) introduction of DNA fragments into a cloning vector (plasmid, cosmid or virus) and (3) introduction of cloned DNA into a bacterium such as *E. coli*. The culture of the host bacterium results in a large number of colonies which carry different pieces of DNA. The colonies can be screened for specific genes (Fig. 12.9). By cutting extracted genomic DNA with two restriction enzymes, a library of random fragments can be produced and it is possible to calculate the probability of including any particular DNA sequence in such a library. If every DNA segment is transformed and maintained with equal efficiency, then the probability (P) of the presence of a desired DNA sequence in a genome bank is as follows:

$$P = 1-[(1-x/L)F]^N$$

where x = length of desired segment of DNA, L= average length of DNA fragment cloned, F= fraction of genome represented by an average fragment, and N= number of transformant colonies containing cloned DNA.

cDNA library can be developed by isolating mRNA and synthesizing the complementary DNA using enzyme reverse transcriptase (Fig. 12.10). This cDNA can be used for cloning purposes after addition of linkers.

Irrespective of the starting material, genomic library of a given organism will be the same but cDNA library, on the other hand, is tissue specific. For example, a human kidney cDNA library will be different from a human liver cDNA library.

Isolation and Purification of DNA from Plant Cells

Due to large size of host DNAs, it is rather difficult to isolate them in intact form. However, isolated DNA fragments of 50 kb or larger size are sufficient for cloning purposes. Isolation of DNA from plant cells becomes difficult because of the rigid cell wall, the presence of vacuoles which release considerable amount of organic acids in the presence of nucleases. For the isolation of DNA from leaf tissues of wheat and barley, the following procedure outlined by Muthukrishnan (1990) is followed:

The leaves are grinded in mortar. The ground leaves are added to a solution comprising proteinase K and EDTA. Solution is then agitated for three hours at 50°C. After three hours, a mixture of chloroform and isoamyl alcohol (24:1)

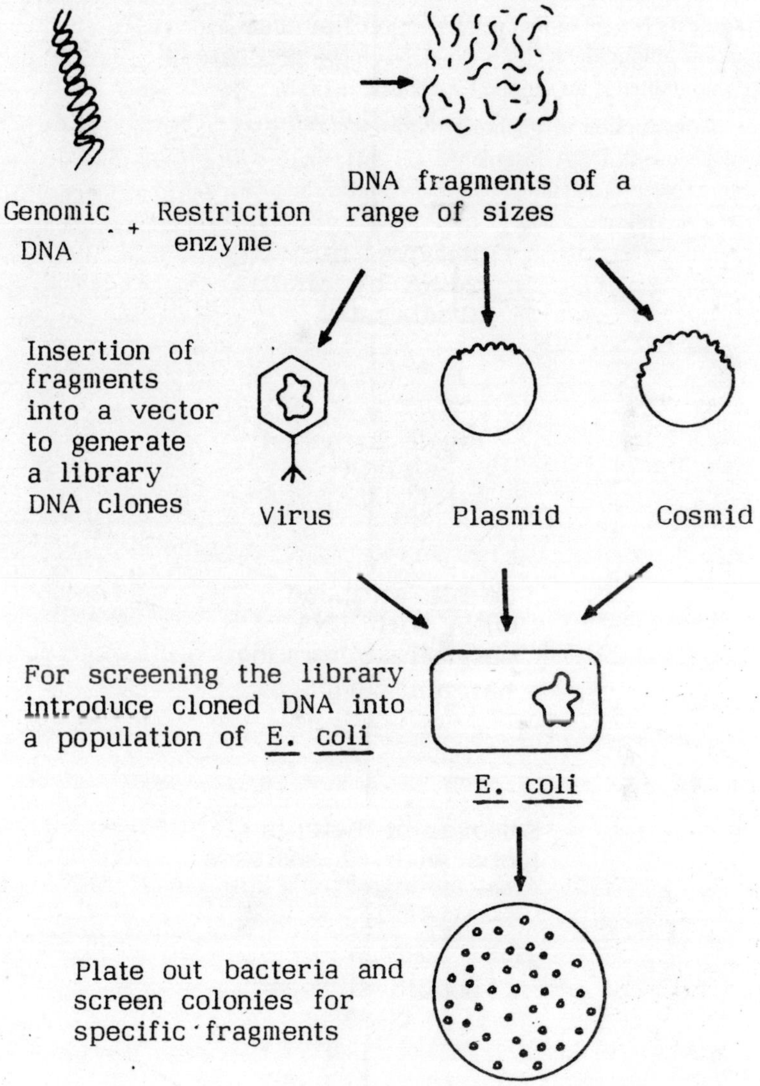

Fig.12.9. Construction of a genomic library.

Isolation of mRNA from tissue

↓ Reverse transcriptase

cDNA copies

↓

Removal of mRNA by alkali treatment

↓

Single stranded cDNA

↓

Double stranded cDNA with DNA polymerase forming hairpin loops

↓

Removal of hairpin loops with S1 nuclease

↓

Double stranded cDNA

↓

Insertion in vector to make library

Fig.12.10. Construction of cDNA library.

is added and flask is agitated at 150 rpm for 5 mnts without heating. Sample is spinned for 10 mnts at 7000 rpm at 10°C. Top layer is then transferred to the dialysis tubing which is kept in dialysis buffer (dialysis buffer is prepared by mixing 1 M Tris, 0.5 M EDTA, 2 M NaCl in distilled water) over night in a cold room with slow stirring. Dialysed samples are then poured into a siliconised graduated cylinder to measure the volume and then to a siliconised Erlenmeyer flask. Cesium gradient is set up for centrifugation. For every 30 ml of sample in the dialysis bag, 27 gm of CsCl is added. Also for every 30 ml of sample, 1.2 ml of ethidium bromide (10 mg/ml) is added. Flasks are then wrapped with aluminium foil to prevent light from nicking the DNA. After the CsCl and ethidium bromide are thoroughly mixed, sample is divided into two 40 ml Beckmann Quick-seal tubes. Balanced pairs are loaded into VTi-50 rotor. Ultracentrifuge is set up at 20°C at 48,000 rpm for 12-16 hours. DNA is then extracted in dark from the tubes with the help of syringe and needle. DNA is again centrifuged at 20°C at 60,000 rpm for 8 hours. DNA bands are then located with the help of ultraviolet light and subsequently collected. This process yields DNA of upto 1 mg of high molecular weight. The length of the extracted DNA fragments varies from 45-60 kb.

DNA SEQUENCING

DNA sequencing involves the determination of the exact order of the four bases or nucleotides on the DNA strand. The task of sequencing DNA chiefly comprises three steps: (i) the cloning and preparation of the DNA to be sequenced, (ii) the sequencing reactions and electrophoretic separations and (iii) the analysis of ensuing data. There are two important methods of DNA sequencing: (1) Chemical method and (2) Chain termination method.

Chemical Method

This method was reported by Maxam and Gilbert (1977). A restriction fragment of DNA is labelled with ^{32}P at either its 5' or 3' using either of the enzymes polynucleotide kinase or terminal transferase. From a restriction map, an enzyme is selected to remove a small piece from one end of the molecule leaving just one end labelled. The DNA is then chemically cleaved at specific residues in five different reactions. These reactions are partially completed and the partial digestion products are separated on a polyacrylamide gel and autoradiographed. The fragments having the labelled terminus are seen. DNA sequencing using chemical method is detailed in figure 12.11.

Chain-termination Method

This method was reported by Sanger *et al.* (1977). In this method also radiolabelled fragments are generated but this is done by copying single-stranded DNA in the presence of a labelled nucleotide with enzyme DNA polymerase. The polymerase also incorporates dideoxynucleotides into the strand. Because of this incorporation, there is no free hydroxyl to accept the next residue resulting in termination of chain. The fragments are analysed in the same procedure as followed in chemical method. This method makes use of the 2', 3'-dideoxy analogs of the deoxyribonucleoside triphosphates in which the 3'-OH function is replaced by a hydrogen atom. Chain termination method needs a primer to be utilized for initiation of DNA synthesis. Usually the DNA to be sequenced is cloned into a phage known as M13. It is filamentous, male specific coliphage which contains a single-stranded DNA genome. DNA sequencing thus provides a detailed insight of a particular fragment of DNA.

RESTRICTION ENDONUCLEASES

These are enzymes which cleave double-stranded DNA into fragments. Restriction enzymes cut double stranded DNA at specific sites. Restriction enzymes are categorised into three groups: Type I, Type II and Type III. Type I and Type III enzymes are non-specific in their cleavage and possess modification (methylation) and ATP, requiring restriction (cleavage) activity in the same protein. As Type I enzymes cleave in a random fashion, they are not suitable for gene manipulation. Type II enzymes are site specific in their cleavage nature, therefore, they have become important tools in the molecular cloning of DNA. Restriction endonucleases have been placed under this category.

Restriction endonucleases cut the double stranded DNA at specific sites resulting either in cohesive termini or blunt-ended DNA. Except Dpn1, all restriction endonucleases identified so far cut unmethylated DNA. Dpn1 cleaves only methylated DNA. It has been observed that those restriction endonucleases which create cohesive ends are more suitable in comparison of those which create blunt ends because the joining of blunt ends by T4-DNA ligae is very inefficient.

The most commonly used restriction endonucleases and their recognition sequences are given in Table 12.1. The best known example of a restriction endonuclease is the enzyme EcoR1. This restriction enzyme attacks the nucleotide sequence GAATTC and makes single-strand cuts called "nicks" between A and G thus opening the circular DNA strand (Fig. 12.12). Any two strands of DNA cut by the same restriction endonuclease unite easily as they possess the same sticky ends e.g. complementary base sequences. Since different endonucleases fragment DNA in different ways, mapping of the genes can be done by comparing the DNA fragments formed by different endonucleases. Such maps are known as restriction maps.

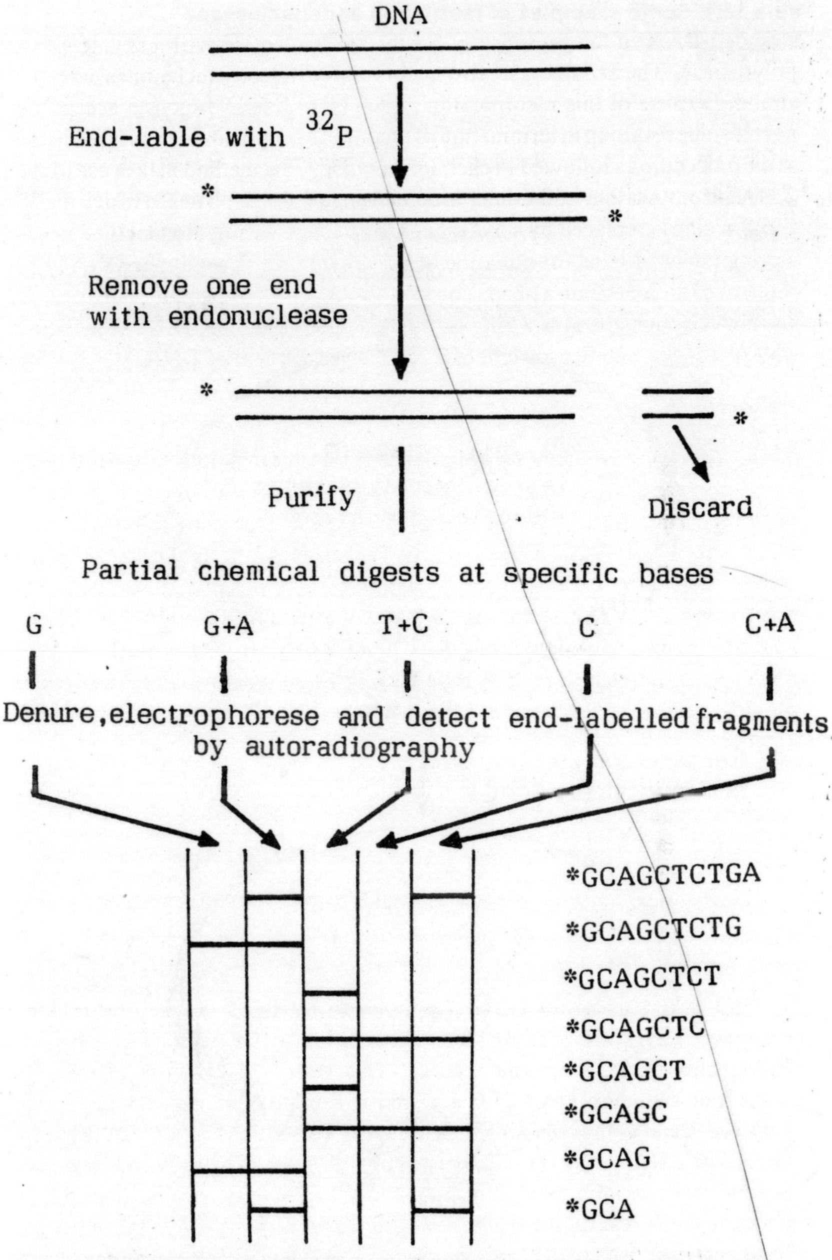

Fig.12.11. DNA sequencing by the Maxam-Gilbert method.

Table 12.1. Some examples of restriction endonucleasese

Enzyme	Source of enzyme	Recognition and cleavage site
EcoR1	*Escherichia coli*	G↓AATTC
Bam H1	*Bacillus amyloliquefaciens*	G↓GATCC
Hind III	*Haemophilus influenzae*	A↓AGCTT
Pst 1	*Providencia stuartii*	CTGCA↓G
Sma 1	*Serratia marcescens*	CCC↓GGG
Sau 3A1	*Staphylococcus aureus*	↓GATC

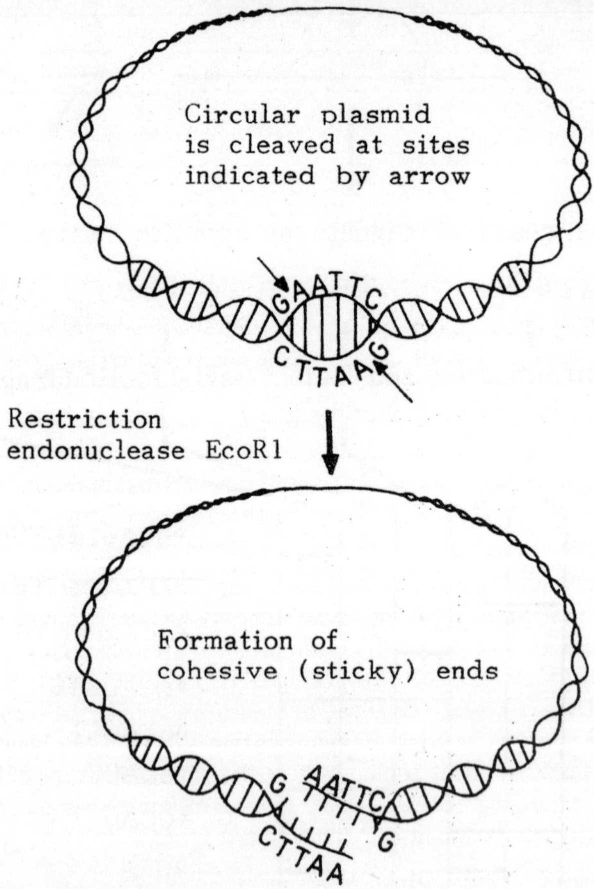

Fig. 12.12. Diagram showing action of EcoR1 on plasmid DNA producing cohesive sticky ends.

Significance of Restriction Endonucleases

Among all the restriction enzymes, restriction endonucleases have immensely contributed towards our understanding of gene structure and function. The DNA fragments produced due to cleavage of small genome by restriction enzymes can be used for wide range of genetic manipulations. The main applications of restriction endonucleases include physical restriction map construction, sequencing of large DNA genomes, construction of gene libraries and molecular cloning.

LIGASE

The enzyme DNA ligase is used to covalently link the 3' -OH terminus of one strand to the 5'-phosphate end of a second strand. Thus the enzyme can repair single-strand breaks such as those present in hydrogen-bonded sticky ends of two DNA preparations that have been cleaved with the same restriction enzyme. DNA ligase from *E. coli* requires NAD as cofactor, while DNA ligase from T_4 and higher organisms utilizes ATP as cofactor. In addition to repairing single-stranded breaks as does *E. coli* enzyme, T_4 ligase can link RNA and DNA strands aligned on a complementary DNA template. Another enzyme RNA ligase joins two single-stranded DNA, or two RNA molecules, or RNA molecule to single-stranded DNA molecule (Sugino *et al.*, 1977).

During the process of ligation, generally the vector anneals back to the original form, which decreases the percentage of real recombinant molecules formed. Linkage of two different DNA molecules is favoured by a high concentration of DNA molecules. Therefore, after ligation of DNA segments at high concentrations of DNA, the mixture is sometimes diluted so that circularization decreases as the length of the DNA segment increases. Due to this reason, short DNA segments are cloned more efficiently than long ones.

Sgaramella *et al.* (1970) reported that T_4 ligase enzyme can join (blunt-ended) fully base-paired DNA duplexes. Although blunt-end joining occurs at a much slower rate as compared to sealing of single-stranded nicks, it is of immense importance especially for joining DNA molecules that lack cohesive ends. DNA segments obtained by shear, by using two different restriction enzymes, or by using restriction enzymes that produce blunt ends, can be joined without adding homopolymer tails. Blunt end ligation of longer molecules obtained by shearing DNA at random rather than at specific sequences offers the possibility of cloning intact genes.

APPLICATIONS OF RECOMBINANT DNA TECHNOLOGY

Recombinant DNA technology, besides providing the material and procedures for gaining insight into vital molecular processes and mechanisms, has opened up new opportunities for highly specific manipulation of the genetic material. Once applied and developed to a sufficient degree, it promises ultimately to provide a powerful additional tool to the plant breeders. This technology broadens the possibilities of transferring genes between unrelated organisms and creating novel genetic information by the specific alteration of cloned genes.

Recombinant DNA technology can be helpful in provding a wide variety of chemicals viz., L-amino acids, organic acids, vitamins, enzymes and antibiotics. Genetic engineering can be profitably employed to enhance the fermentation capabilities and other beneficial properties of the microbes.

Transfer of specific genes, form one species to another may be of great signficance in exploiting diseases, insect and pest resistance mechanisms more efficiently. Recombinant DNA technology can also be helpful in development of nitrogen fixing cultivars in non-leguminous crops, development of increased photosynthetic efficiency in crop plants and improvement of nutritional quality of crop plants.

13

BIOTECHNOLOGY IN MEDICINE AND HUMAN HEALTH

The recent results of recombinant DNA technology have revolutionized the field of clinical medicine. This has greatly helped in isolation of rare biological molecules, preparation of wide range of vaccines, development of highly specific diagnostic laboratory tests and prenatal diagnosis of human genetic diseases. The medicinal and scientific approaches to human health involves two main aspects: (a) curative and (b) preventive. Biotechnological achievements can be applied towards preventive and in some cases curing human diseases. Pharmaceutical industries were the first to use recombinant DNA technology for improving existing processes or the elaboration of substances that were not usually commercialised due to their limited production e.g. human peptide hormones. Some of the products which have been produced using biotechnological processes are detailed below:

Human Growth Hormone

Human growth hormone (HGH; somatotropin) is used in the treatment of dwarfism caused by pituitary malfunction. Infants with a deficiency of growth hormone becomes dwarf whereas those who produce too much become giant sized. Until the advent of genetic engineering, the only source of HGH was human pituitary glands.

Human growth hormone is a single polypeptide of 191 amino acids. A large precursor of the protein is synthesized in the somatotrophs of the enterior pituitary and the mature form is secreted in the circulatory system. Growth hormone induces liver and perhaps other cells also to produce other hormones called somatomedins which are insuline-like factors that stimulate proliferation of mesodermal tissues such as bone, cartilage and muscle.

Goeddel *et al.* (1979) obtained expression of mature human growth hormone by *Escherischia coli*. They isolated a specific restriction fragment of about 550 base-pairs having the coding sequence for amino acids 24-191 (COOH terminus) of HGH. Later on construction of a synthetic DNA molecule corresponding to the first 24 amino acids of mature HGH and an ATG initiation codon were made. These two fragments were ligated and inserted into an expression vector and the resultant plasmid gave rise to the production of 2 mg of HGH per litre of culture.

Commercial Production of HGH: The growth hormone coding sequence is initially cloned into a bacterial expression plasmid using part of a cloned growth hormone cDNA and several synthetic oligonucleotides. Expression of human growth hormone in *E. coli* involves five steps: (1) several overlapping, complementary oligonucleotides are synthesized and ligated together. One strand of the resulting small DNA fragment contains the coding sequence for the first 24 amino acids of the mature human growth hormone, (2) a recombinant DNA plasmid with a full length of human growth hormone cDNA (which is not expressed) is cleaved with restriction enzymes that release fragments containing the complete growth hormone coding sequence after codon 24, (3) the synthetic fragment and the partial cDNA- containing fragment are ligated together to yield a new fragment containing the complete coding sequence of mature human growth hormone, (4) the new fragment is ligated into a restriction site just downstream from the lactose promoter-operator region cloned on a plasmid and (5) the resulting recombinant DNA plasmid is introduced into bacteria in which synthesis of human growth hormone can be induced with IPTG (isopropylthiogalactoside), an inducer of lactose operon.

2. Interferons

Interferons are glycoproteins having molecular weight of about 20000. The interferons are classified into three groups: (i) IFN -α,(ii) IFN- ß and (iii) IFN-γ).These are species specific cytokinins which have in common an inhibitory action upon the multiplication of viruses in cells more or less adjacent to the affected ones. The bulk of the α-interferon is obatined from buffy coats (leukocyte pellets). The leukocytes can be brought into suspension (10^7 cells/ml) and induced to secrete interferon with Newcastle-disease or Sendai virus. This process may also be carriedout in serum-free media which has advantages in the subsequent purification process. The producers of ß-interferon are diploid fibroblasts, particularly the FS-4 cell line whereas γ-interferon are produced by B- or T- lymphocytes. It has been reported that IFN-γ is 50 to 100 times more effective than the other interferons in destroying tumor cells. The American Cetus Corporation has improved the activity of the human ß interferon simply by substituting one amino acid (serine) for one of the cysteine residues. Japanese companies have succeeded in producing cloned γ-interferon. Interferons are being used for the treatment of several diseases including a rare form of cancer called hairy cell leukamia.

3. Insulin

Insulin, a small protein hormone, is used for controlling diabetes. Bacterially produced human insulin is now available in the market. This compares well with the classical insulin that was extracted from slaughtered animals. Products of microbial origin are often safer than those derived from traditional sources. Human insulin produced by microbes, for example, does not cause the allergic reaction associated with commercial insulin obtained from its traditional source, cows and pigs. Genetic engineering techniques have found one of the most important uses for the production of insulin (Fig. 13.1).

4. Vaccines

Vaccine is a preparation which contains an antigen composed of whole disease causing organisms (attenuated) or parts of such oraginsms. Vaccines are used to confer immunity against the antigen. The vaccines currently in use have three types of antigens:

1. Dead microorganisms: Vaccines with bacterial or viral agents inactivated by various methods in order to preserve their antigen capacity.
2. Live or attenuated organisms: Living vaccines developed through exhaustive selection of mutant strains.
3. Toxoids: Some bacteria exert their pathogenic effects through the secretion of toxic proteins that act in various tissues of the host. These proteins are transformed into toxoids in order to make them lose their toxic activity but still retain their antigenic specificity. These toxoids induce the production of antibodies capable of neutralising the native toxins by blocking their active sites.

Recombinant DNA technology can be used to clone the gene for the protective antigen protein. Small peptides may be synthesised, conjugated to a carrier molecule and utilised as the sole antigen source for eliciting a protective immune response. Recombinant DNA technology has yielded remarkable results in the development of antimalarial vaccine. A number of vaccines against viruses have been developed using recombinant DNA technology (Table 13.1).

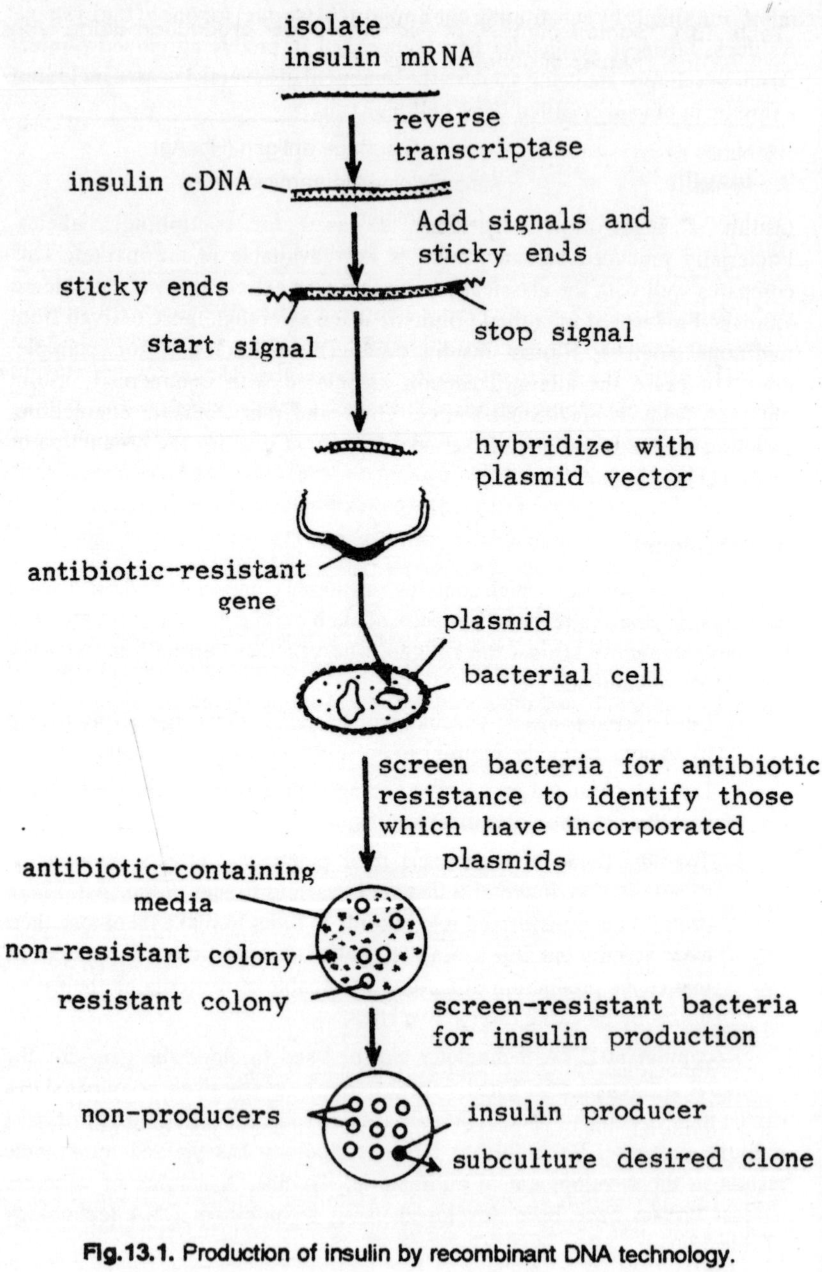

Fig.13.1. Production of insulin by recombinant DNA technology.

Table 13.1. Some examples of vaccines under production using techniques of genetiuc engineering

Virus	Cloned gene
Hepatitis B	Hepatitis B surface antigen (HbsAg)
Influenza	Hemaglutinine/Neuraminidase
HIV (AIDS virus)	Surface antigen
Herpes	Various coat subunits

Vaccines may also be developed using recombinant vaccinia virus genome. In this method DNA for surface antigen epitopes from viruses such as Influenza A or Hepatitis B are cloned into the vaccinia viral genome to the vaccinia derived and recombinant antigens and mounts a protective immune response to them. This results in a polyvalent vaccine.

Through recombinant DNA techniques, it has become possible to identify the DNA sequences that code pathogenecity determinants e.g. production of toxins. By eliminating those specific sequences, it is possible to transform the microorganism into an attenuated variant that retains all other characteristics and can be used as an immunizing agent.

Vaccines can be elaborated against the surface structure of the microorganism. Those structures, usually proteins, are extremely important in the initial stages of infection, and are specific for each microorganism. Those proteins can be produced on large scale by DNA fragments cloned in *E. coli* having the sequence coding adhesion structures.

5. Monoclonal Antibodies

Antibodies are specific proteins produced by the immune system of higher animals and human beings as part of the immune response to the presence of a specific antigen. Monoclonal antibodies, on the other hand, are highly specific antibodies derived from only one clone of a specific hybridoma cell and therefore exactly of the same type. They recognise only one specific site of an antigen.

Monoclonal antibodies can be produced using hybridoma technique (a method of creating pure and uniform antibodies against a specific antigen). The hybridoma technology is greatly helping the development of effective vaccines against human, animal and also plant diseases.

The preparation of monoclonal antibodies is outlined in Fig. 13.2. A mouse is repeatedly immunized with the desired antigen. As a result, there is proliferation of B lymphocytes. These lymphocytes are removed from the mouse and are fused with myeloma cells by incubating the mixture in the

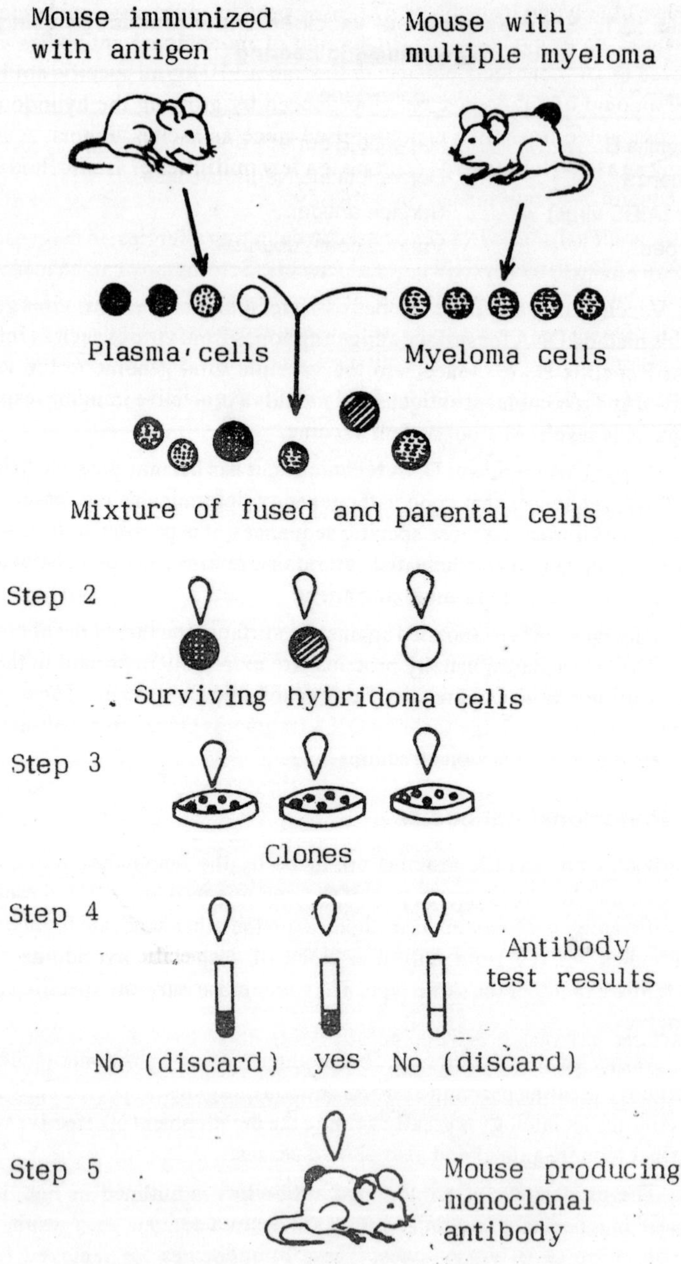

Fig. 13.2. Method of producing monoclonal antibodies.

presence of polyethylene glycol. The fused product gives pure and homogeneous bodies on a regular basis. Production of monoclonal antibodies can be achieved *in vitro* but the yield is very low (about 10 μg/ml specific antibody). Large amount of antibodies can be produced by growing the hybridomas in histocompatible or immunocompromised mice as ascites tumors. A mouse bearing an ascitic tumor may accumulate a few millilitres of ascitic fluid which may contain 5-20 mg/ml specific antibody.

Monoclonal antibodies can be used in detecting allergies, to diagnose viral diseases and to detect certain types of cancers. Serotherapy can be made more effective with the administration of a monoclonal antibody. Monoclonal antibody technology can provide sufficient quantities of standardised immune reagents for diagnosis of plant diseases and epidemiology. Productivity of plants can be enhanced through the development of specific antibodies to different strains of plant pathogens.

6. Antibiotics

Antibiotics are chemical substances, produced by microorganisms and synthetically, that has the capacity in dilute solutions to inhibit the growth of, and even to destroy, bacteria and other microorganisms. Most antibiotics are produced during the resting phase of microbial growth, as a result of a type of metabolism, called secondary metabolism, which is different from that occurring during growth and division and is concerned with maintaining the cell during this phase of its life.

Among the metabolites of interest to the pharmaceutical industry, the antibiotics are the most important and the most challenging from the point of view of applying recombinant DNA technology. The application of recomibinant DNA technology in transferring the genes involved in bioconversions into an antibiotic producing organism reduces a number of steps of cumbersome chemical extraction. Of the almost 5,000 antibiotics known, only about 100 have been marketed, of which 70% are derived from *Streptomyces*. However, the fungal produced ß-lactam antibiotics, penicillins and cephalosporins contribute to about 25% of the world market. ß-lactam antibiotics produced by eubacteria have been discovered and their derivatives establishes a major market share alongwith ß-lactam produced by actinomycetes. Highly productive strains already exist and significant improvements in existing manufacturing procedures for producing cephalosporin antibiotics could be acheived through the expression of one or two foreign genes in *Cephalosporium acremonium*. A number of antibiotics are being produced using microorganisms as source materials. Some useful antibiotics which are being produced on commercial scale employing biotechnological techniques are listed in table 13.2.

Table 13.2. Some important antibiotics

Antibiotic	*Source*	*Spectrum of activity*
Cephalosporin	*Cephalosporium* species	Gram-positive bacteria
Chloramphenicol	*Streptomyces venezuelae*	Gram-positive and gram-negative bacteria
Cycloheximide	*Streptomyces griseus*	Fungi
Erythromycin	*S. erythreus*	Gram-negative bacteria
Gentamycin	*Micromonospora*	Gram-negative bacteria
Griseofulvin	*S. griseus*	Fungi
Kanamycin	*S. kanamyceticus*	Broad spectrum
Lincomycin	*S. linocolnensis*	Staphylococci
Neomycin B	*S. fradiae*	Mycobacteria
Oligomycin	*S. diastatochromogenes*	Fungi
Oxytetracycline	*S. rimosus*	Broad spectrum
Penicillin G	*Penicillium chrysogenum*	Gram-positive bacteria
Polymixin B	*Bacillus polymyxa*	Gram-positive bacteria
Streptomycin	*Streptomyces griseus*	Gram-negative bacteria

Among all the antibiotics, penicillin (Fig. 13.3) is the most important antibiotic medically. One of the most effective and widely used penicillins is penicillin G obtained from *Penicillium chrysogenum*. Penicillin is a secondary metabolite, produced after the period of rapid logarithmic growth of the organism. Penicillin fermentations and production of most other antibiotics require 7-8 days for high yield. The production of the various semisynthetic penicillins constitutes an interesting combination of micobiological and chemical syntheses. The microbiological conversion of penicillin G into 6-aminopenicilic acid (6-APA) has fewer steps and is cheaper than the chemical conversion. In the production of semisynthetic penicillins, 6 APA forms a chemical nucleus to which side chains can be attached, yielding new antibiotics (Fig.13.4).

After penicillin, the next useful antibiotic discovered is streptomycin.

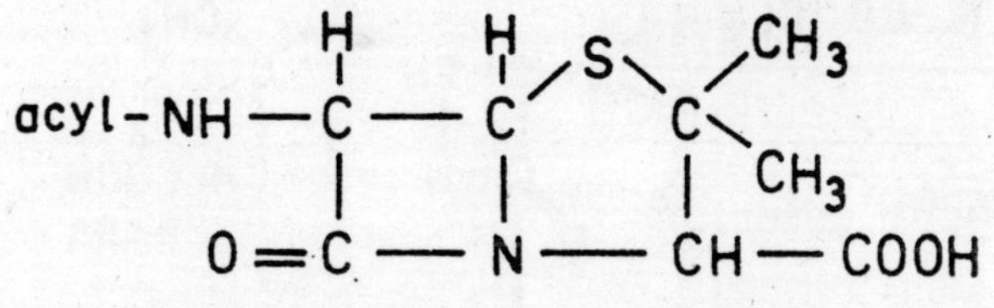

Fig.13.3. General structure of penicillin.

Sterptomycin is produced by the filamentous bacterium *Streptomyces griseus*. It is a trisaccharide which blocks bacterial protein synthesis and impairs membrane function. Streptomycin production follows the normal pattern of secondary metabolic synthesis, taking place immediately after exponential phase. Ion-exchange resins have provided the most effective method of separating antibiotic from components of the medium and the various minor antibiotics from streptomycin itself.

7. Laboratorial Diagnosis

Modern medical practice depends on laboratory tests for the diagnosis of many diseases. Recombinant DNA technology allows for the production of highly specific diagnostic tests.

1. Infectious Disease Diagnosis

One of the most serious aspects of the infectious diseases is their capacity to get rapidly disseminated in the community. Some of the common infectious diseases are cholera, small pox, measles, meningitis, and hepatitis. These diseases lead to serious damage to human health. Infectious disease diagnosis mainly depends upon isolation and identification of pathogen. This may take several days. The development of diagnostic test to identify pathogenic organisms by the identification of organism-specific DNA sequences have provided rapid and specific diagnosis. One of the characteristics being used for the identification of viruses, bacteria, or other agents is their surface structure. Antisera prepared against surface components can be utilised in the serological typing of such organisms.

Recombinant DNA methods could be used to isolate specific antibodies to surface antigens of microorganisms. These specific antibodies could be used in diagnostic tests. The production of monoclonal antibodies by hybridoma technique against specific antigenic determinants of the cellular surface provides an efficient method for the correct identificaiton of several microorganisms.

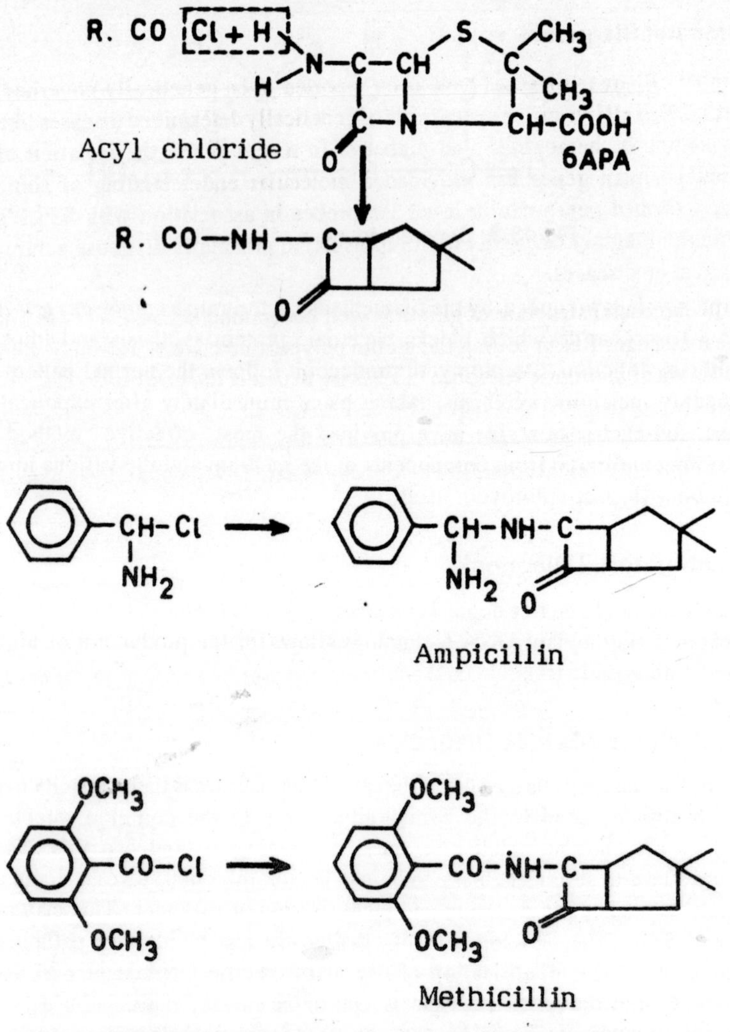

Fig. 13.4. Production of semisynthetic penicillins.

Among the bacterial diseases, the infectious diarrhoeas have been extensively studied from the point of view of development of new identificaiton techniques, especially those caused by enterotoxigenic *E. coli* (ETEC). ETEC strains are very difficult to identify in clinical laboratories because conventional methods detect enterotoxin production by animal or cell culture assays which are very complicated in routine work. Techniques of genetic engineering have greatly contributed to the development of alternative assays for the detection of ETEC.

2. Prenatal Diagnosis

A number of human diseases have been reported to be genetically governed. About 1-2% of all live births are carriers of genetically determined diseases like phenylketonuria, hemophilia and diabetes. In recent years, the isolation of individual human genes has provided a molecular understanding of some diseases. Cloned genes can be used as probes in association with RFLP's (Restriction Fragment Length Polymorphisms) to prenatally diagnose a number of genetic diseases.

The molecular structure of globin is well understood because it is easy to isolate messenger RNAs coding the globin polypeptide chains. This helped in construction of appropriate probes, necessary to detect the position of restriction sites in globin gene DNA sequences. DNA analysis of hemoglobinopathies (caused due to defects of hemoglobins) was then possible and this is one of the best examples of clinical applications of recombinant DNA technology applied to prenatal diagnosis.

14

BIOTECHNOLOGY IN AGRICULTURE

During the last decade, tremendous progress has been made in the area of plant cell, tissue and organ culture. *In vitro* techniques constitute an important component of biotechnology, and have the potential not only to improve the existing cultivars, but also for the synthesis of novel plants and early release of high-yielding plants resistant to various diseases, pests, stresses and temperature. Green revolution of recent years has been the result of efforts that led to the improvement and better selection of species of agronomic interest. Some specialists feel that biotechnology is the second phase of green revolution. Studies are already in progress to obtain crop plants having higher photosynthetic efficiency, improved fixation of atmospheric nitrogen, increased nutritional quality and greater tolerance of salinity, alkalinity, acidity, deficiency of nutrients and other stresses. Wide hybridization through embryo and ovule culture both at the interspecific and intergeneric levels has been achieved in a number of crops through the culture of the hybrid ovules and embryos. *In vitro* fertilization techniques have also been employed to over come incompatibility Haploids produced through anther culture and their diploidization have resulted in the release of varieties in crops such as wheat and rice. Some of the areas where biotechnology has played a significant role are detailed below.

1. Production of Pathogen-free Plants

Plants traditionally being vegetatively propagated are systemically infecte with viruses and other pathogens which greatly reduce yield and also quality o the marketed commodity. Virus diseases like potato leaf roll virus (PLRV) o potato virus Y (PVY) for example, can cause upto 95% reductions in the tube yield of potato crops. Increase in yield upto 300% (Averaging 30%) has bee reported following replacement of virus-infected stock with specific pathogen free plants (Murashige, 1980). Eradication of viruses and other pathogens i extremely important in order to optimize the yields and also facilitate th movement of living materials to other countries of the world. The species for which virus-free plants have been obtained using tissue culture techniques are listed in table 14.1. Since majority of viruses infect plants in a systemic manner, their elimination may be achieved through meristem tip culture. This method consists of culturing shoot tips of infected plants *in vitro* which then give rise to virus free plants.

The meristem-tip culture method has its major application in vegetatively propagated species where parental clones have become virus infected. There are two distinct advantages of meristem-tip culture: (a) The integrity of the original plant or clone is maintained with minimum genetic variation, (b) Methods of meristem-tip culture are relatively straight forward and can be used directly with slight modification for clonal propagation of the species. By using shoot meristem culture, specific pathogen-free plants or "clean stock" have been recovered from diseased plants in several economically important crops viz., sugarcane, potato, citrus. Chrysanthemum, narcissus, carnations and other ornamental crops freed of viruses produce increased number of blooms per plant and improved size of flowers. Recently, banana cultivation has received major setback because of the spread of two viral diseases; bunchy top and infectious chlorosis. Virus free plants have been obtained in banana through meristem tip culture (Fig. 14.1).

Table 14.1. Plant species for which virus-free plants have been obtained

Species	*Virus eliminated*
Allium sativum (Liliaceae)	Garlic mosaic virus
Brassica oleracea(Brassicaceae)	Cabbage black ringspot Virus, Turnip mosaic virus, Cauliflower mosaic virus
Chrysanthemum sp. (Asteraceae)	Chlo rotic mottle, Green flower, Stunt, vein mottle, Virus B
Dahlia sp. (Asteraceae)	Dahlia mosaic virus, Tomato Aspermy, Vein mottle, Virus B
Glycine max (Fabaceae)	Soybean mosaic virus
Ipomoea batatas (Convolvulaceae)	Feathery mottle, Hanmon *Mosaic*, Rugosa mosaic
Lilium sp. (Liliaceae)	Cucumber mosaic virus
Manihot sp. (Euphorbiaceae)	African Cassava mosaic
Musa sp. (Musaceae)	Cucumber mosaic virus
Nicotiana tabacum(Solanaceae)	Tobacco mosaic virus
Saccharum officina rum(Poaceae)	Mosaic
Solanum tuberosum (Solanaceae)	Potato virus
Zingiber officinale (Zingiberaceae)	Mosaic

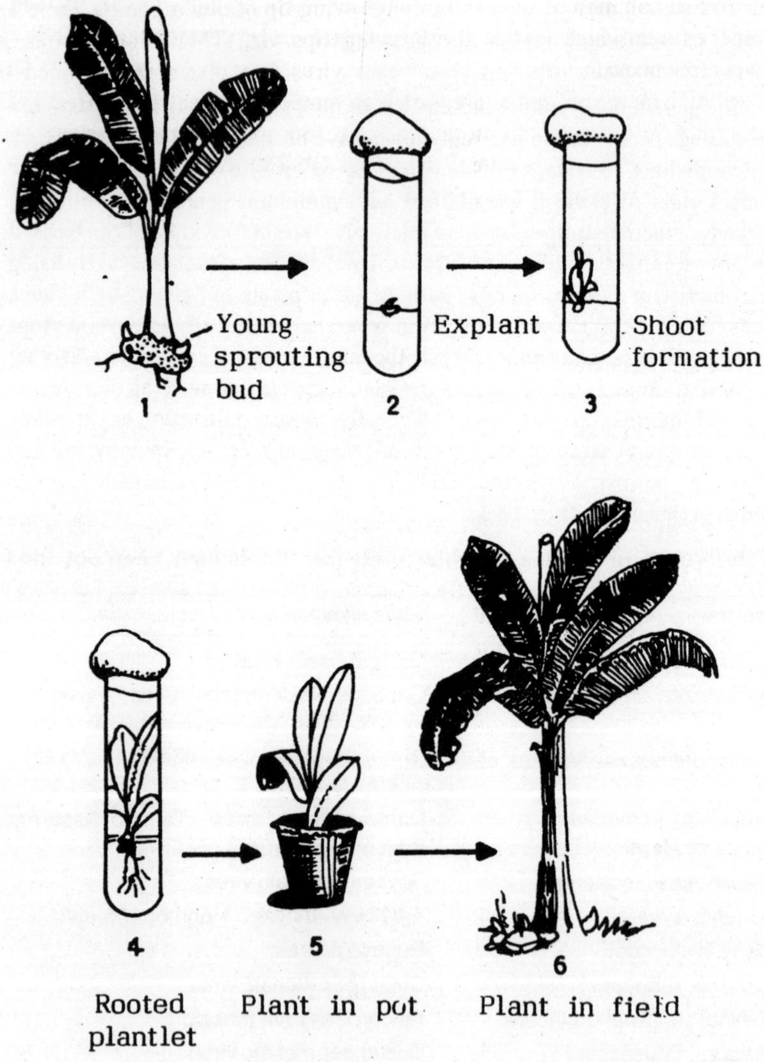

Fig.14.1. Schematic representation of shoot tip culture method to obtain virus free plants of banana.

Although the apical meristems are often free of viruses, this can not be regarded as a phenomenon of universal occurrence. The success of meristem-tip culture depends on several factors. One of the most important factors is the relative distribution of viruses in the growing tip of donor plants. There are some viruses which invade the growing tips viz., TMV, Potato virus X, Cucumber mosaic virus. In such cases virus free plants are obtained by combining meristem-tip culture with high temperature treatments. Heat treatment can be given to the mother plant before excising the meristem tips. Eradication of Potato Spindle Tuber Virus (PSTV) requires two consecutive cycles of heat treatment followed by meristem-tip culture.

2. Production of Disease Resistant Plants

This is one of the most useful applications of tissue culture in crop improvement. In potato, somaclones have been screened for both late and early blight resistance. The parent, Russet Burbank, is highly susceptible to both early and late blight. From among more than 800 plants, about 2% of the somaclones displayed enhanced resistance to late blight (*Phytophthora infestans*) which was transmitted through subsequent tuber generations. Several disease resistant somaclones have been recorded from a population of 500 plants screened for resistance to early blight (*Alternaria solani*). In maize, somaclonal variation has induced resistance to race T of southern corn leaf blight, *Drechslera maydis*. In sugarcane, resistance to diseases like fiji and downy mildew have been recovered. In lucerne, selection of cell lines and plants resistant to the toxin of *Fusarium oxysporum* has been accomplished.

3. Improvement of Nutritional Quality

Once sufficient variation exists in the breeding material, selection of the desired genotype becomes necessary. Selection can be made either in field conditions or in tissue cultures. Selections in tissue cultures may be done at cell, callus or plantlet levels. In comparison of conventional field selections, tissue culture method is more efficient as a large number of plantlets can be grown in a small space and also manipulation of culture medium may provide a rapid method for screening of callus cells.

One of the major sources of protein for human and animal consumption is constituted by the proteins contained in seeds of many plant species (Table 14.2). The cereals and legumes which are major sources of storage seed proteins, contain limited amount of certain amino acids which are essential for human beings. Majority of these cereals are deficient in lysine whereas legumes are deficient in sulphur amino acids. A wide range of approaches have been employed for improving nutritional quality of various crop plants. Important among them are selecting cell lines resistant to amino acid or their analogues of lysine, tryptophane, proline and phenylalanine. Isolation of variants over-producing specific amino acids in culture has been successful, but expression in the whole plant and especially in the seed has not yet reached the level required to make an impact on protein quality. An increase in free threonine levels has been achieved in barley and maize seeds. In rice, on the other hand, selection for lysine over producers led to genotypes with increased seed lysine and protein, but decreased yields. Alteration in the relative proportions of different proteins in the seed may be a more effective way of manipulating amino acid balance. The genes which code for a number of storage proteins have been cloned. The genes encoding the french bean protein, phaseolein, has been expressed in sunflower.

Table 14. 2. Seed storage proteins of some cereals and legumes

Species	*Proteins*	*Protein type*	*Approximate proportion of total protein %*
Cereals			
Barley(*Hordeum vulgare*)	Hordein	Prolamin	55
Maize(*Zea mays*)	Zein	Prolamin	52
Oat(*Avena sativa*)	Endosperm globulin	Globulin	56
Rice(*Oryza sativa*)	r-Globulin	Globulin	10
	Glutelin	Glutelin	80
Wheat (*Triticum aestivum*)	Gliadin	Prolamin	40
	Glutelin	Glutelin	
Legumes			
Pea (*Pisum sativum*)	Legumin (11s) Vicillin (7s)	Globulin	80
French bean (*Phaseolus vulgaris*)	Phaseolin (G1)	Globulin	50
Soybean(*Glycine max*)	Glycinin (11s) B-Conglycinin (7s)	Globulin	60-80

4. Selection for Salt and Drought Tolerance

Continued efforts to increase intensity of cropping for increasing production from limited land resource by extending irrigation facilities have resulted in the gradual build-up of salt concentrations in the soil. This has resulted in loss of productivity on such soils. The production of selections with increased salt and drought tolerance has relied on obtaining variants arising during cell cultures.

Salt tolerant lines have been produced in crop palnts such as tobacco, tomato, cereals. The cultivated tomato is highly sensitive to saline conditions but some populations of *Lycopersicon minor* grow in littoral habitats where level of salinity is very high. Plants of *L. esculentum* when subjected to half strength sea water collapsed. Rush and Epstein (1976) determined that resistance was associated with an ability of cells to cope with increased sodium levels. Rush and Epstein (1980) observed that the salt tolerant *L. cheesmanii* accumulates sodium in leaf and petiole in amounts exceeding 20% of the total dry weight but in *L. esculentum* sodium tends to be excluded under salt stress and is lethal at concentrations above 5%. Responses of protoplast and callus cultures to increased salinity levels have revealed that cell cultures under salt stress are frequently associated with accumulation of proline. Salt tolerance has been incorporated into rice lines with improved plant type. Some of the improved rice cultivars viz., IR 42, IR43 and IR52 are salt tolerant.

In *Nicotiana sylvestris* and *Capsicum annuum* it has become possible to select salt tolerant lines which can grow in 0.34 M NaCl. In *Agrostis stolonifer* cell lines with high tolerance to copper and zinc have been developed. The plants regenerated from these lines show similar level of tolerance to copper and zinc. Salt tolerant cell lines in alfalfa have been selected which are resistant to 1% NaCl. The selected cell line grew poorly in the absence of salt. Plants have been regenerated from salt tolerant cell lines.

Development of cultivars tolerant to drought can contribute significantly in agriculture economy. Tolerance to drought is a polygenic trait and involves highly complex osmo-regulatory functions. In tissue cultures, simulated drought conditions have been achieved through incorporation of non-penetrating osmotic solutes such as PEG (polyethylene glycol) and dextrans in the media. In tomato, water stress tolerant lines have been successfully isolated by using PEG as selective agent.

5. Production of Genetically Variable Plants

The success of any crop improvement programme depends on the usable genetic variability in the base population. Cells in culture offer an excellent system for inducing variation and regenerating pure mutant types. Genetic variation can be an option to lessen our reliance on cost-intensive germ plasm collection and conservation programmes. Variant producing capacity of cell culture can be augmented to a great extent by employing physical and chemical mutagens. Somaclonal variation has been extensively exploited for the improvement of asexually propagated crops viz., potato, sugarcane. From cell cultures, some superior cultivars have already been produced in sugarcane which are high yielding, drought resistant and temperature tolerant. Similarly, superior cultivars of rice, tobacco and mustard have also been developed using somaclones (for details see chapter 4).

6. Biofertilization

Molecular nitrogen in the atmosphere is converted to biologically usable forms by nitrogen fixing microorganisms e.g. *Rhizobium*. The most sophisticated approach to biofertilization is to create plants that possess the genetic capacity for nitrogen fixation. Attempts are being made to transfer genes for nitrogen fixation (*nif* genes) from bacteria to plants. *Rhizobium* species carry *nif* genes on large 500 kbp plasmids. There are 17 *nif* genes carried in eight, closely linked transcription units or operons. Dixon and Postgate (1971) successfully transferred the *nif* genes to free living nitrogen fixer *Klebsiella pneumoniae.* Later on they transferred these genes to *Escherichia coli* converting it into a nitrogen fixer.

7. Rapid Clonal Propagation

Tissue culture has found its best commercial application in production of cloned plants at a very high rate as compared to conventional methods. It is important specially for initial building up of propagation stock of elite clones or individual plants which are otherwise slow to multiply. A number of agriculturally important plants have been clonally multiplied and propagated (for details see chapter 8).

8. Germplasm Storage

The primitive cultivars and wild relatives of crop plants constitute a pool of genetic diversity which is invaluable for future breeding programmes. There are over 20,000 plant species which are rare or threatened with degradation of their natural habitats. The most economical form of storing germ plasm for seed propagated species is as seeds. However, there are certain limitations of this method: (a) loss of viability of seeds with the passage of time, (b) destruction of seeds by seed-borne pathogens, insects and pests, (c) discrete clones can not be maintained except for apomictic species and (d) it can not be used for vegetatively propagated crops. Various methods of *in vitro* storage of germplasm are thus of great practical significance for long term storage of germplasm. In cultures plants can be maintained by serial subculturing but this method involves the risks of loss of plant material due to microbial contamination, accidental loss through labelling or media errors and is uneconomical also. Presently, there are two main approaches to *in vitro* germplasm storage: slow growth technique and cryopreservation.

1. Slow Growth Techniques

Germplasms can be stored in cultures at low temperatures (4-8° C) with reduced light (an 8 hour day). At low temperatures the ageing of the plant material is slowed down but not completely stopped as observed in freeze-preservation. Vegetative growth of shoot cultures can be slowed down by (i) using growth retardants, (ii) inducing osmotic stress (by adding 4-8% mannitol in the normal culture medium) and (iii) reducing the supply of carbon source. By using slow growth cultures, interval between two subcultures may be prolonged. The technique of cold storage has been successfully used for a wide range of plant species (Table 14.3).

Tabel 14.3. Some examples of *in vitro* storage of plants

Species	*Period in storage*	*Viability percentage*
Fragaria virginiana	72 months	100
Fragaria vesca	72 months	100
Lolium multiflorum	10-11 months	80-100
Malus domestica	12 months	100
Medicago sativa	15-18 months	94-95
Rubus sp.	14 months	95
Trifolium repens	15-18 months	89-92

2. Cryopreservation

Cryopreservation (Greek, kryos= frost) is literally 'preservation in the frozen state . In practice, this is generally taken to mean storage: at very low temperatures: above solid carbon dioxide (-79° C), in low temperature deep freezers (-80° C or below, in vapour phase (approximately -150° C) or in liquid nitrogen (-196°C). The chief advantage of storage at very low temperatures like these is to slow down considerably, or even to halt, metabolic processes and biological deterioration. In addition, cryopreserved materials remain genetically stable. The freeze-storage of plant cells offers a wide range of possibilities in agriculture by offering new technology for the conservation of germplasm of rare, elite and other crop species.

The principle underlying cryopreservation basically involves bringing the culture to a state of non-dividing and zero metabolism, which is achieved by subjecting it to superlow temperatures in the presence of cryoprotectants. Entire plants have been regenerated from cells, meristems and embryo frozen and stored in liquid nitrogen for varied lengths of time (Table 14.4).

Table 14.4. Some examples of freeze-preservation of shoot tips/embryo/ plantlets at -196° C

Species	*Organ*	*Storage period*	*Survival percentage*
Carrot (*Daucus carota*)	Somatic embryos	60 days	100
Cassava (*Manihot esculentum*)	Shoot tip	Not known	21
Chick pea (*Cicer arietinum*)	Shoot tip	3 weeks	27-36
Pea (*Pisum sativum*)	Shoot tip	26 weeks	61
Peanut (*Arachis hypogea*)	Shoot tip	3 weeks	23-31
Potato(*Solanum tuberosum*)	Shoot tip	5 min	42-76
Tobacco (*Nicotiana tabacum*)	Pollen embryos	Not known	2-31
Tomato (*Lycopersicon esculentum*)	Shoot tip	5 min	30-45

Freeze preservation of living plant cells involves four steps: (a) freezing, (b) storage, (c) thawing and (d) reculture. The steps involved in the freeze preservation for *in vitro* cultures are outlined in figure 14.2. The cryobility of cell cultures depends on a number of factors viz., morphological and physiological conditions of the plant material, cryoprotective agents, rate of freezing, storage temperature, method of thawing and the culture conditions. Most cultures require special conditions to present maximum freeze tolerance. Usually a rapid growth rate is required in order to produce cells with comparatively small vacuoles. Low water content in the cells minimizes the damage due to formation of ice crystals and promotes the chances of the material surviving freezing and thawing. Various chemicals such as glycerol, dimethylsulphoxide (DMSO), proline, mannitol/sorbitol, sucrose, glucose and polyethylene glycol are added to culture media before freezing. These compounds are called cryoprotectants. Cryoprotectants protect against ice damage and depress both the freezing point (=melting point) and the super cooling point of water and elevate the divitrification point. Freezing of the plant material is done either by slow cooling or rapid cooling. Frozen materials are stored in liquid nitrogen (-196°C). Rapid thawing is done to avoid the damaging effects of ice crystal growth. The thawed material is washed several times to remove the cryoprotectant and recultured to regenerate plants.

9. Biological Control of Agricultural Pests

Insects consume nearly one-third of human food supplies on earth. One of the major goals of biotechnology is to develop target specific biological pesticides

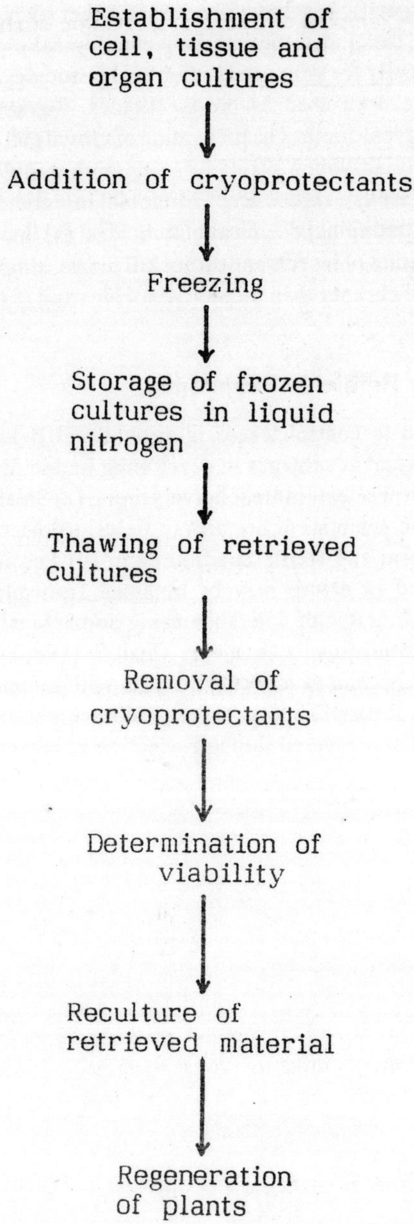

Fig.14.2. Sequential steps in cryopreservation.

that will kill some specific pest but will not harm other species. Several insect-specific pathogens, developed through biotechnological processes, are being produced commercially for their use as microbial pesticides. ***Bacillus thuringiensis*** inocula have been used as bioinsecticides. Toxin produced by the bacterium kills the gypsy moth. The prevention of crown gall disease formation is achieved by spraying young plants or seeds with avirulent species *Agrobacterium radiobacter* var. *radiobacter*. Microbial insecticides have a number of advantages over traditional chemical insecticides: (a) they are specific for a small number of species of insects and do not kill plants, animals and beneficial insects, (b) they are cheaper than organic pesticides and (c) they do not leave any residue effect.

10. Selection for Herbicide Resistance

Herbicides are used to control weeds in agricultural fields. Tissue culture techniques offer several advantages in developing herbicide - resistant plants : (1) The techniques for selection are relatively rapid, (2) Small volumes of cells represent the genetic potential of hectares of fields, (3) Metabolic studies are facilitated by uniform and sterile condition and (4) Possible new types of resistance not found in nature may be obtained from plant cell cultures. Attempts to produce herbicide tolerance using somaclonal variation and *in vitro* selection have already been made. Chaleff (1980) has reported the selection of several ***Nicotiana tabacum*** mutants with increased tolerance to herbicide picloram. Resistance to paraquat, chlorosulphuron and sulphometuron methyl has been recovered through selection in tobacco cell culture.

15

INDUSTRY AND FERMENTATIONS

Recent advances in molecular biology have considerably expanded the scope of using the microbes in new areas of industry, agriculture, medicine and food production.Three major factors have caused a tremendous revitalization and diversification of the fermentation industry. The first reason is the continuing development of the recombinant DNA technology started in the early 1970s. This has led to an entirely new market of microbially produced recombinant DNA products such as vaccines, polypeptide hormones, insulin, interferons, bulk enzymes, growth hormones, amino acids and immunoregulators.The second major factor is reshaping the fermentation industry due to the development of animal and plant tissue culture systems. Both of these culture systems require fermentation conditions that are much different from those required for traditional fermentation of heterotrophic bacteria. Plant tissue fermentation has importance in the continuing development of the basic physiology of crop plants, germ plasm preservation, genetic engineering and plant secondary metabolites. Moreover, the regeneration of plants from tissue cultures will require new fermentation technology in order to deal with regenerated plant fidelity or trueness, the combined mutation, protoplast fusion and selection processes, and actual production of plant biochemicals in fermenters. Animal tissue culture fermentation has now become extremely important, mostly because of the developement of hybridoma technology. The final force driving the modernization of fermentation is the development of sophisticated computer-control systems. Computer control of fermentation has also opened the doors to new types of fermentation that are feedback-controlled, so that the reactions of the organisms themselves actually run the process (Strohl and Pfister, 1986).

BIOREACTORS

A bioreactor (fermenter) is a device in which substances are treated to stimulate biochemical transformation by living cells or by *in vitro* cellular components such as enzymes. In other words, in a bioreactor, the transformation of raw materials into desired products is carried out by the enzyme systems of living organisms or by isolated enzymes. Bioreactors are used for fermentations and enzyme transformations. The former employ living cells whereas the latter use cell free systems. In enzyme reactors, substrate transformation is brought about by employing immobilized enzymes.

The environment maintained in a fermenter can be divided into three broad groups: biological, chemical and physical. The biological environment is favourable to a biological process when only desired organisms, contributing to the process, are present. The chemical environment involves the composition of the microbial growth medium, which should contain the desired concentration of substrate or microbiological nutrients and synthetic precursors, free from inhibitory substances and maintained at the desired pH. The physical environment refers to the temperature of the system.

Reactor Design

Apart from the need to ensure sterile operation, three requirements dominate fermenter design: good mixing, adequate oxygen transfer and heat removal. The fermenters are of two types: open and close. The open fermenters allow continuous processing with substrate entering at one end and products leaving at the other end. In closed fermenters the processing is done in batches. This type of fermenter is used for the production of antibiotics. Most fermentations to produce useful products are done in the aerated stirred vessels (Fig. 15.1).The stirred-tank fermenter is a versatile design and is used in a range of sizes from one-litre laboratory units to production-scale vessels of typically 100-tonne capacity. The fermenter is usually constructed as an upright cylinder, with dished ends to facilitate liquid drainage during medium discharge at the bottom, and of splashes and disrupted foam during operation at the top. The vessel is constructed of a suitable grade of stainless steel because such fermenters are resistant to steam sterilization. Majority of fermenters are fitted with stirring gear which helps in breaking clumps of mycelial organisms. Flat vertical paddles are used as agitator. Two disadvantages of the stirred vessel system are the high energy requirement for agitation and the damaging effect on cells of the shear effect. Airlift bioreactors provide a significant alternative approach to conventional stirred reactors (Fig. 15.2). Airlift bioreactors are favoured because of three reasons: (a) savings in power costs, (b) reasonable oxygen transfer rates and (c) lower shear due to hydrostatic agitation. The airlift bioreactors are of two types: (1) bioreactor with an outer column and an inner cylinder referred to as the draught tube and (2) bioreactor with an external downflow section.The airlift principle has been particularly useful in the

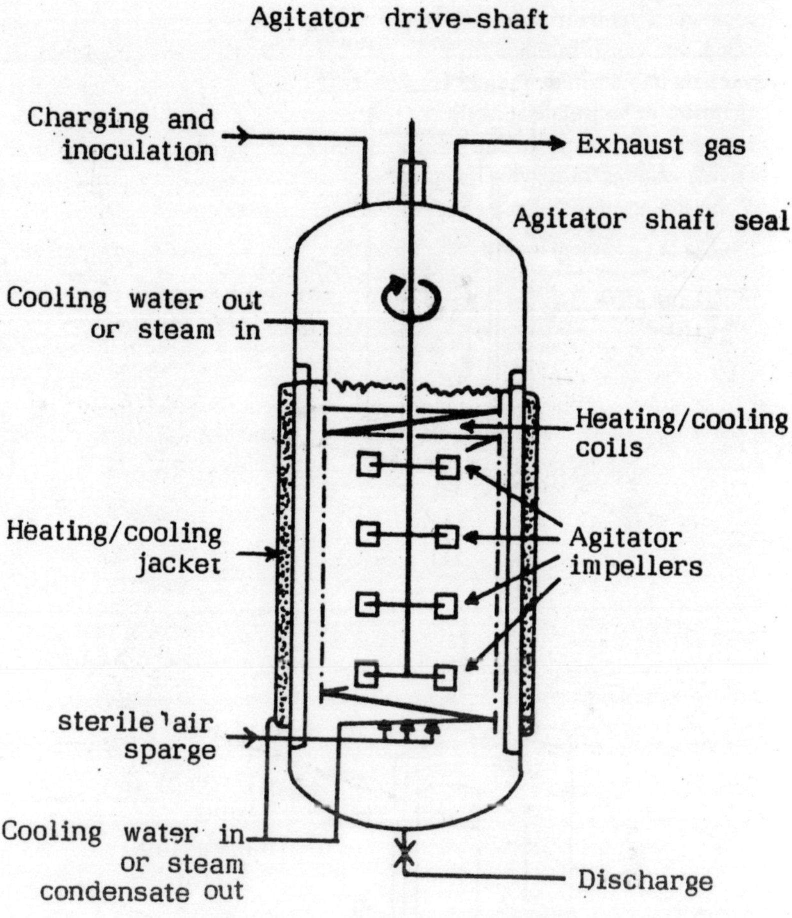

Fig.15.1. Stirred-tank fermenter.

process for production of single-cell protein.

FERMENTATION

Fermentation may be defined as a process by which products are obtained from organisms cultured in contained system. Fermentation consists of four components: the fermenter hardware or vessels, the sensors that monitor the system, the medium and the organisms. Recently one more component is added to the fermentation process i.e. computer control system.

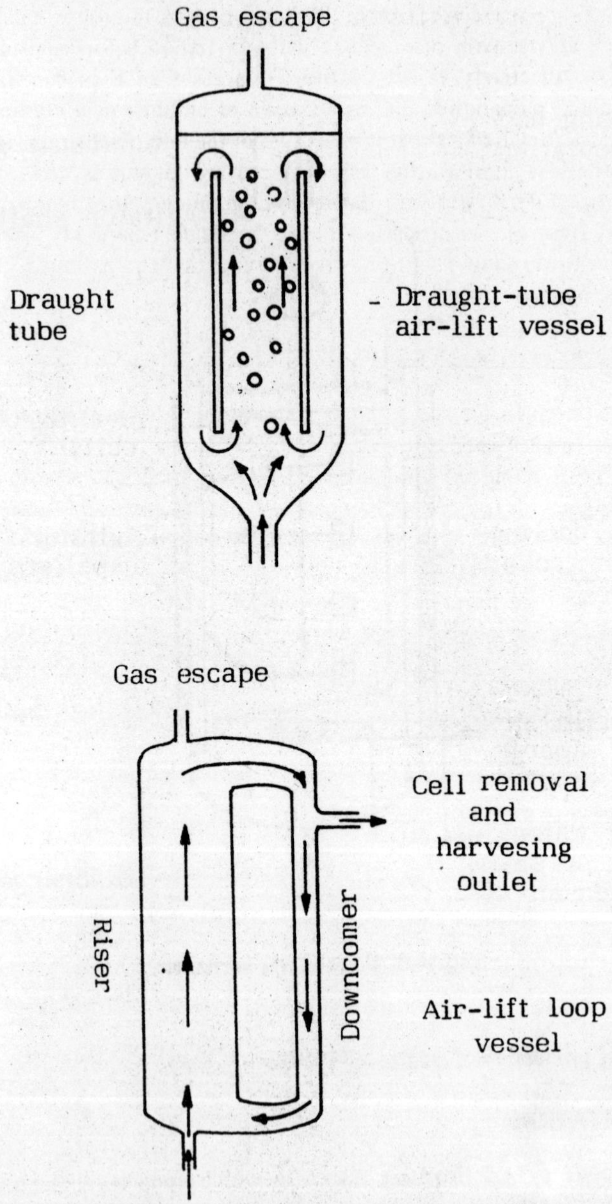

Fig.15.2. Basic types of air-lift bioreactors.

Fermentation processes can be divided into two broad groups: upstream processes and downstream processes. All the operation before starting the fermenter are collectively called upstream processes such as sterilization of the fermenter, preparation and sterilization of culture medium and the preparation and growth of a suitable inoculum of the microbial strain. All the operations after the fermentation has occurred are known as downstream processes (Fig. 15.3). After the fermentation, the product must be recovered and purified from the fermentation broth. Methods commonly used are distillation, centrifugation, filtration, solvent extraction and various forms of chromatography.

FERMENTATION INDUSTRY

The fermentation industry has its ancient times with the natural production of wine from grapes carriedout by anaerobic culture of grapes with the indigenous yeasts. It was not until the late 1800s, however, that Pasteur proved that fermentation process was due to the action of the yeasts. Once this link was established, a burgeoning fermentation industry was started for the production of acetone, butanol and glycerol needed for world war I. Since then varied industrial products are being manufactured using fermentation technology.

Alcoholic Beverages

Majority of alcoholic beverages are manufactured by metabolic activity of *Saccharomyces cerevisae* and other yeasts. Some of the important

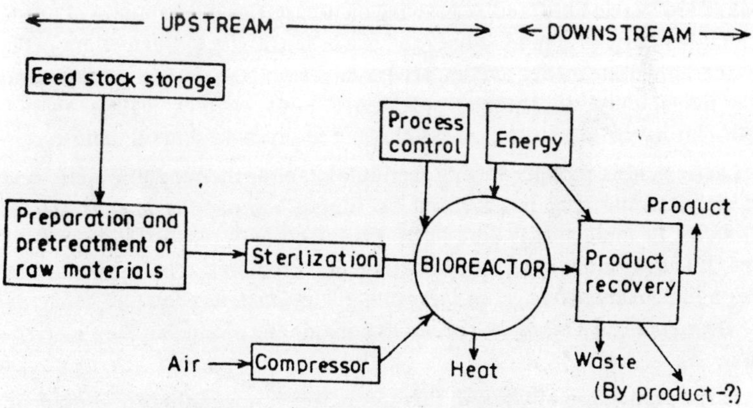

Fig.15.3. A flow chart of upstream and downstream processes.

alcoholic beverages and materials fermented for their production are presented in table 15.1. Modern fermenters generate a vigorous natural movement of the yeast and the fermentation is faster and there are less chances of microbial contamination. These fermentation vessels are equipped with automatic cleaning and sterilizing equipments. The biotechnological techniques have greatly helped in manifold increase in total production of alcoholic beverages.

Table 15.1. Alcoholic beverages and production substrates

Alcoholic beverages	*Materials fermented*
Beer	Barley and other cereals
Brandy	Grapewine,apple, cherry
Gin	Grain mash, flavoured with juniper berries, anise etc.
Rum	Sugarcane juice, mollases
Vodka	Cereal, potato
Whisky	Cereals

Organic Acids

Citric acid is produced on commercial scale from fermenting molasses or purified glucose syrup from maize using *Aspergillus niger* as the microorganism. This fungus also produces gluconic acid. Citric acid and isocitric acids are produced by yeasts like Candida growing on n-paraffins. Some strains of yeasts elicit higher citric acid productivity and lesser sensitivity to fluctuations in the crude carbohydrate media, as compared to *Aspergillus niger*. Citric acid is used in the manufacture of shampoo, soft drinks, jams, candies, leather tanning, electroplating and in clearing metal-clogged in pipes used in oil industry.

Lactic acid is produced using thermotolerant or thermophilic lactic acid bacteria. Itaconic acid is produced by fungus *Aspergillus terrens* and *A. itaconicus*. Itaconic acid is used in the manufacture of plastics and synthetic fibres. It is also sometimes used in detergents. Propionic acid is produced by bacteria like *Propionobacterium shermanii*, *Clostridium propionicum*, *Bacteroides ruminicola* and *Veilonella* species as a major end product of their metabolism.

In the production of vinegar, two successive fermentation are carriedout. Addition of yeast, *Saccharomyces diastaticus* to the fermentation maximizes the carbohydrate assimilation. Second stage fermentation is carriedout by *Acetobacter* species which converts ethanol to acetic acid.

Amino Acids

Commercial production of a number of amino acids is being done using fermentation technology. Out of 20 naturally-occurring amino acids, 1[illegible] are produced by industrial fermentation or enzyme conversion. Amino acids are used to supplement foodstuffs and for the treatment of diet dificiency. Amino acids or their derivatives are also used in the manufacture of leathers, cosmetics, biocides and surface active agents. L-glutamic acid is produced by the fermentation activity of bacterium *Corynebacterium glutamicum*. L-proline and L-lysine production involves auxotrophic bacteria such as *Brevibacterium flavum*.

Amino acids are produced in conventional stirred tank reactors. The commercial production of L-isoleucine is accomplished by the addition of α-aminobutyrate to the culture medium. Other examples of precursor addition include the production of L-tryptophan by *Hansenula anomala* growing on glucose, which has been improved by the addition of anthranilic acid to the medium.

Vitamins

Vitamins can be synthesized by chemical processes or by biological means. Riboflavin is produced from the fungus *Eremothecium ashbyii* whereas vitamin B_{12} is cheifly synthesized from *Streptomyces griseus*. *Streptomyces* also produces antibiotics. During fermentation process, change in the operational conditions may lead to the production of vitamin rather than antibiotics.

Polysaccharides

Synthesis of polysaccharides through chemical means is expensive but polysaccharides produced through biological means are cheaper. Among the industrial polysaccharides, dextrans and xanthans are being produced by bacteria *Leuconostoc mesenteroides* and *Xanthomonas campestrans* respectively. Dextrans are used to expand the volume of blood plasma for transfusion. Xanthans are safe blood additives.

Polysaccharides have many industrial uses due to their capacity to alter the rheological properties of aqueous solutions. Large-scale industrial extraction of starch alginates, and to a lesser extent xanthan, provides raw materials for the detergent, paper, textile, adhesive, paint, food and pharmaceutical industries.

Enzymes

Several thousand enzymes possessing different substrate specificities are now known. The bulk of enzymes used commercially are obtained from microbial sources, and are produced by fermentation processes. Enzymes from microbial, rather than plant or animal, sources are usually used because of their easier availability, greater stability and their variety and ease of genetic manipulation. The industrial enzymes are produced from different species of bacteria and fungi (Table 15.2). These enzymes are used in industries like baking, brewing, distilling, textile and paper. The use of microbial enzymes in either isolated or immobilized form to modify certain chemicals could play an important rôle in the pharmaceutical industry. Immobilization can be defined as the process whereby the movement of enzymes, cells, organelles, etc. in space is completely or severely restricted, usually resulting in a water-insoluble form of the enzyme, although water soluble immobilized enzymes can be formed by linking the enzyme to soluble polymers such as dextrans. Immobilized enzymes are also sometimes referred to as bound, insolubilized, supported or matrix-linked enzymes. Immobilized enzymes offer certain advantages over soluble enzymes. These include (1) higher stability, (2) easier recovery of the products, (3) possibility of repeated use and (4) less chance of contamination.

Techniques of immobilization: There are six different techniques which are used for immobilization of biocatalysts, namely (1) entrapment in polymer matrixes, (2) adsorption on a carrier surface, (3) encapsulation inside semipermeable membranes, (4) covalent coupling to a carrier surface, (5) crosslinking between enzyme molecules (copolymerization), (6) ultrafiltration. Covalent bonding is the most widely used method for immobilizing enzymes.

Table 15.2. Some common examples of industrial enzymes

Enzyme	*Source*	*Uses*
α-amylase	*Aspergillus oryzae*	Baking, brewing
	Bacillus licheniformis	Paper and textile
Cellulase	*Aspergillus* species *Trichoder-marusei Penicillium* species	Fruit and vegetable processing
Lectase	*Saccharomyces* species	Lactose hydrolysis
Pectinase	*Aspergillus niger*	Extraction and clarifica-tion of fruit juices
	Bacillus licheniformis	Detergent and leather industry

Mass Culture of Plant Cells

Despite major advances in the field of synthetic organic chemistry, the plant species still continue to contribute significantly in both quantity and product range of chemicals used by a number of industries including food, pharmaceutical and agrochemicals (Tables 10.1, 10.2). A wide range of industrial products can be extracted from *in vitro* cultures. With the developement of new techniques for the isolation and selection of cell lines, there have occurred enhancements in yield of a number of secondary compounds to values above those found in whole or parts of whole, plants (Table 10.3).

Plant cells have now been grown in batch, semicontinuous and continuous culture. Batch culture has been the traditional way of growing plant cells in liquid culture. Almost all the semicontinuous culture work has been essentially batch culture, in which large volumes of culture are removed from the culture vessel and replaced with the same amount of fresh nutrient. Table 15.3 lists some of the bioreactors that have been designed, constructed and used for a wide range of experiments and applications.

Table 15.3. Bioreactors used for growing plant cells

Plant	*Reactor type*				
	SS	*ST*	*BC*	*AL*	*GL*
Acer		+			+
Catharanthus				+	+
Coleus	+	+			+
Daucus	+	+			
Digitalis	+	+			
Ginkgo	+		+	+	
Glycine		+	+		
Lithospermum	+	+			
Lolium	+			+	
Morinda				+	
Nicotiana	+	+	+		
Phaseolus		+			
Triticum			+		

SS= stainless steel, ST= stirred tank, BC= bubble column, AL= airlift loop, GL= glass

Although a wide range of microbial and plant cells have been grown in culture but the metabolic rate of plant cells is much lower as compared to microorganisms. However, the range and amount of secondary compounds formed in cultured plant cells are usually greater than in microbial cells. Plant tissue culture biotransformation suffers from lack of an economy of scale because plant cells grow very slowly, with typical generation times of 15 to 35 hours, whereas bacteria double every 18 to 25 minutes. Table 15.4 compares microbial and plant cells in relation to growth in mass cultures.

Table 15.4. Differences between microbial and plant cells

Characteristics	*Microbial cell*	*Plant cell*
Size	2μm³	>10⁵μm³
Doubling time	1 to a few hours	1 to a few days
Individual cells	Frequent	Occasional
Clumps	Rare	Frequent
Colonies	Frequent	Rare
Shear stress	Insensitive	Sensitive
Inoculation density	Small	5-20% of total culture
Variability with in a single culture	Stable	Can vary greatly
Developmental morphology	Spores/pseudomycelium	Organogenesis/ embryogenesis
Product formation	Into the medium	Into vacuole

Biotransformation

Biotransformation is a biological process in which an organic compound is modified into some useful, recoverable products by enzymes contained in the cells. Biotransformations are mostly stereospecific, involving the addition or removal of single chemical groups (by hydroxylation, glycosylation, acetylation and methylation), such as the C_{12} hydroxylation of digitoxin to digoxin, a heart drug, by cultured cells of the foxglove *Digitalis lanata*. Major types of transformations reported in plant cell cultures are summarised in table 15.5.

Table 15.5. Biotransformations which can be achieved by plant cell cultures (Mantell *et al.*, 1985)

Reactions	*Substrate*	*Product*
Reduction	C=C	CH_2-CH_2
	C=C-CO	CH_2-CH_2-CO
	CO	CHOH
	CHO	CH_2OH

Oxidation	CH3	CH_2OH or COOH
	CH_2OH	CHO
	CHOH	CO
	=S	S=O
Hydroxylation	CH	C-OH
	CH_2	CH-OH
	NH_2	NH-OH
Epoxidation	CH=CH	-HC-CH- O
Glycosylation	OH	O-glucose
	COOH	COO-glucose
	CH	C-glucose
	N	N^+-arabinose
Esterification	OH	O-palmitate
	COOH	COO-malate
	NH	N-acetate
Methylation and	OH	O-CH_3
demethylation	N	N^+-CH_3
	N^+-CH_3	= NH
Isomerization	trans	cis
	dextra	Laevo-rotation
	β-OH	α-OH

Biotransformation has been utilised on a massive industrial scale in the steroid industry. Several species of bacteria (e.g. *Acetobacter*, *Gluconobacter*) and fungi (e.g. *Rhizopus*) have been used for the past many years to accomplish the specific alterations in the phenanthrene ring structure and side-chain composition of several plant-derived steroids, e.g. diosgenin, to produce the steroid sex hormones and anabolics. The efficiency of the biotransformation processes may be improved by applying recombinant DNA techniques and microbial screening.

Single-Cell Protein

Single-cell protein (SCP) is microbial biomass meant for human or animal consumption. Yeast cultivated specially as food and feed represents one of the most promising aspects of using microbial biomass as food. The often quoted comparison that a 0.5 t bullock synthesizes less than 0.5 kg of protein every 24 hr, but 0.5 t of soyabeans produces the equivalent of 40 kg protein every 24 hr and 0.5 t yeast generates 50 t in this time, illustrates well the main advantage of microbial SCP. Both photosynthetic and nonphotosynthetic microbes, grown on various carbon and energy sources, are used in fermentation process for the SCP production (Table 15.6).

Table 15.6. Properties and protein contents of some microbes important in production of SCP (after Litchfield, 1983).

Organism	*Carbon and* energy sources	*Crude protein* (%)
Candida lipolytica	n-alkanes	65
C. utilis	Ethanol, sulphite waste liquor	50
Cellulomonas	Bagasse	85
Cephalosporium eichhorniae	Cassava starch	50
Chaetomium cellulolyticum	Agricultural and forestry wastes	45
Kluyveromyces fragilis	Cheese whey (lactose)	50
Methylophilus methylotrophus	Methanol	70
Paecilomyces varioti	Sulphite waste liquour	55
Rhodopseudomonas capsulata	Industrial wastes, sunlight	60
Saccharomyces cerevisiae	Molasses	50
Scenedesmus acutus	CO_2, sunlight	55
Spirulina maxima	CO_2, $NaHCO_3$, sunlight	60

The most useful single cell proteins furnish all essential amino acids. Bacteria and yeasts in several cases have been found to be better protein sources than higher plants which are deficient in amino acid lysine. Bacteria are widely used for SCP production because of their fast growth rates and their ability to utilise a wide range of carbon and energy sources. Production of SCP requires little space and is largely independent of climate. Algae and photosynthetic prokaryotes are potentially inexpensive sources of single cell protein. Algae like *Spirulina*, *Scenedesmus*, and *Chlorella* are being used in dried forms as human food. The inhabitants of the northern shores of Lake Chad in Africa and Aztec Indians of the Mexico habitually eat *Spirulina* which is an important source of SCP because of its high nutritive value and also because, unlike unicellular algae, the filaments of this alga can be relatively easily harvested from water surface by means of fine mesh nets or by simple filtration. SCP is also being produced from bacteria and yeasts like *Saccharomyces*, *Torulopsis* and *Candida*. Fungal mycelium (e.g. *Fusarium*) is also being used for microbial SCP production. Fungal mycelium has the texture of animal meat. Production of SCP is of special interest to human societies that have plentiful supplies of cheap carbohydrate and severe shortage of vitamin and protein -rich food.

Ethanol Production

Due to limited availability of nonrenewable fossil fuels, production of ethanol from renewable plant resources has become important. Ethanol is manufactured synthetically from petrochemicals or biologically through the process of fermentation by microorganisms. Although economically produced worldwide by catalytic hydration of the ethylene from petroleum, industrial ethanol obtained through yeast fermentation of sugar is becoming competitive in the international market mainly as a consequence of rising oil prices.

Ethanol is both a high grade fuel as well as an excellent raw material for the chemical and plastic industries. Production of alcohol involves fermentation of a number of raw materials (Table 15.7). Different crops can be used as raw material for ethanol production. Most plant fibres contain cellulose, hemicellulose and lignin. The cellulose is resistant to hydrolysis because cellulose fibres generally consist of a highly ordered crystalline structure of cellulose surrounded by a lignin seal which acts as a physical barrier to easy hydrolysis. Hydrolysis of hemicellulose yields a mixture of sugars. With cellulosic biomass as a starting material several pretreatments (physical, chemical and enzymatic) are done (Figs. 15.4, 15.5). Average yields of ethanol vary from crop to crop. In sugarcane, average yield of ethanol is 4,000 litres per hectares whereas it is 2,500 litres per hectare in millets. Some other crops or their residues which can be utilised as substrate for ethanol production are banana, maize, paddy and sorghum.

Brazil for many years has used ethanol as automotive fuel in mixture with gasoline and recently as hydrate alcohol. India is world's third largest producer of fermentation ethanol, first being USA and Brazil. In India, over 100 distillaries with an installed capacity of about 700 million litres per anuum are engaged in production of ethanol. Molasses is the most widely used substrate for ethanol production. The average yield of ethanol per ton of molasses is approximately 225 litres.

Industrial quality ethanol, may be produced either by continuous fermentations or in batch cultures. Normally distillate of 65-70% ethanol is collected for production of spirit drinks. Anhydrous ethanol, required for many industrial purposes and for blending into petroleum spirit as mother fuel, is produced by further distillation from benzene.

Recombinant DNA technology and cell fusion technique offer new oppor-

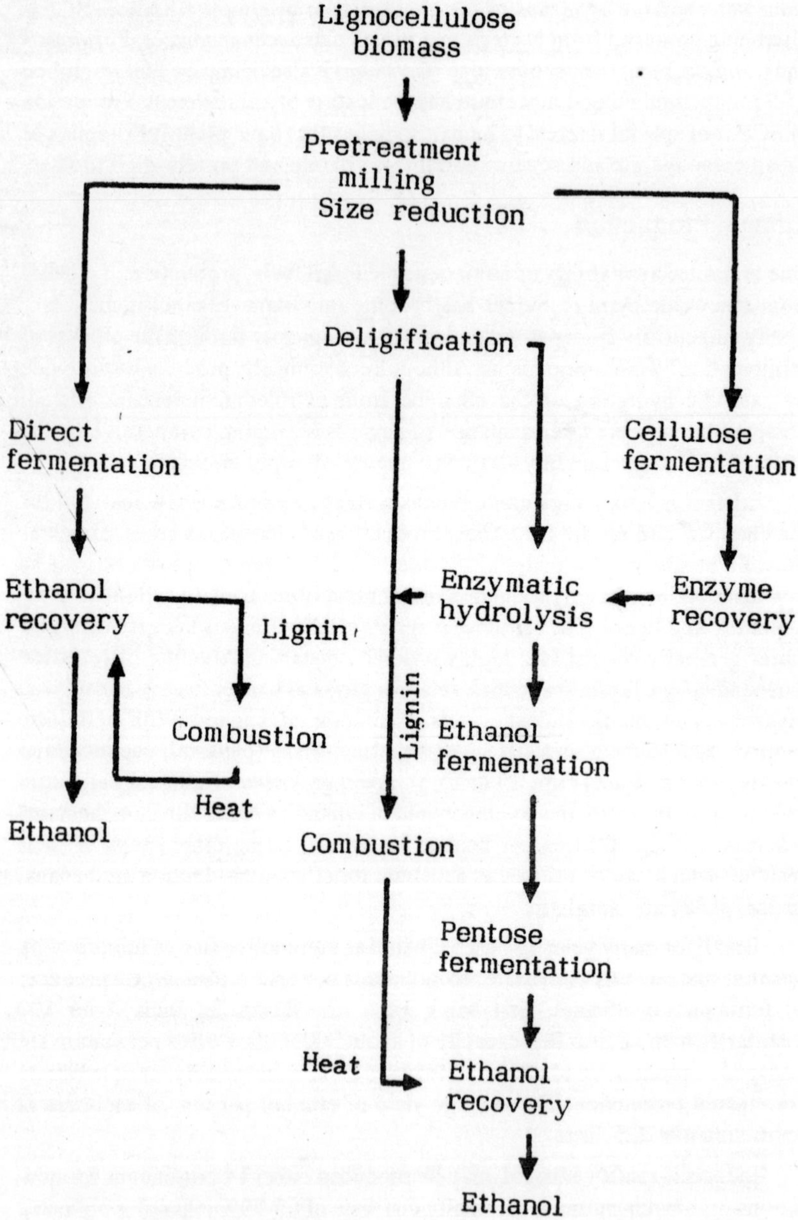

Fig.15.4. Steps in production of ethanol (Acid hydrolysis method)

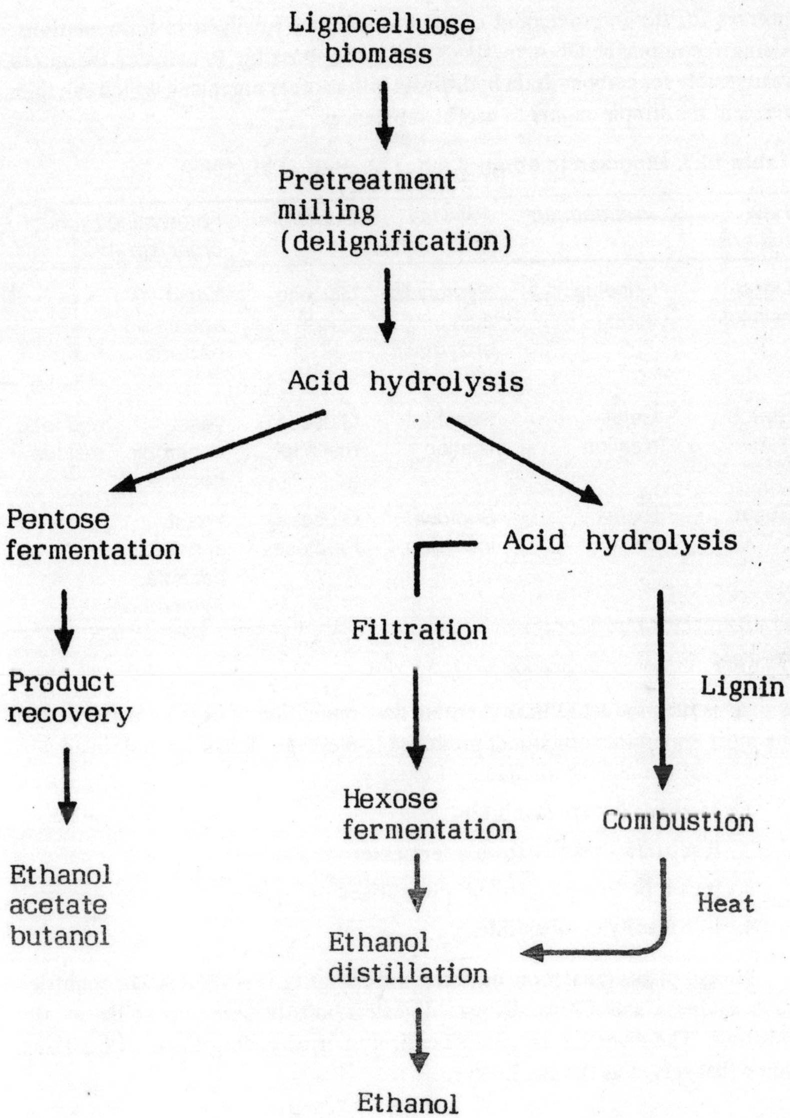

Fig.15.5. Steps in production of ethanol (Enzymatic hydrolysis method).

tunities for the improvement of microorganisms involved in fermentations. Using recombinant DNA methods, it is now possible to transfer the genes responsible for carbohydrate hydrolysis into another organism, which can then ferment the simple sugars to useful substances.

Table 15.7. Biomass to ethanol (after Mantell *et al.*, 1985)

Raw material	*Pretreatment*	*Process*	*Substrate*	*Fermenting organism*	*Product*
Ligno-cellulose	Grinding, milling	Hydrolysis: acid or enzymatic	Glucose xylose	Yeast, anaerobic bacteria	Fuel ethanol
Starch	Gelati-nization	Sacchari-fication	Glucose, maltose	yeast, anaerobic bacteria	and feed residue
Sugar	Extra-ction	Sucrose inversion	Glucose, Fructose	Yeast, anaerobic bacteria, *Zymomonas*	

Biogas

Biogas is the product of the anaerobic decomposition of organic material and the most important component produced is methane. There are certain advantages in using methane as source of energy:

1. Methane is very insoluble.
2. It separates readily from a fermenter system.
3. It can be easily collected, pressurised or liquefied for storage and
4. It is readily combustible.

Biogas plants (anaerobic digesters)are operative in several Asian countries such as India and China. Biogas digesters mostly use crop stalks as the substrate. The digesters are almost entirely buried underground with a fixed dome that serves as the gas holder.

Methanogens

Methanogens are group of bacteria that release methane as a metabolic product. Some of the common methanogens and the substrates on which they act are listed in table 15.8. Methanobacteria occur in the intestine of animals, soil, black pond muds, faeces and anaerobic sewage tanks. Methanobacteria are autorophs and use only CO_2 as the source of carbon.

Table 15.8. Some examples of methanogens and the substrate on which they act

Bacterium	*Substrate*
Methanobacterium formicicum	Formate or hydrogen
M. ruminanticum	Formate or hydrogen
M. mobilis	Formate or hydrogen
M. propionicum	Propionate
M. sohngenii	Acetate, Butyrate
M. suboxydans	Butyrate, Valerate, Caproate
Methanobacillus omelianskii	Hydrgen, ethanol
Methanosarcina barkeri	Hydrogen, CO_2, Methanol and Acetate

Methane Production

In India, China and African countries biogas is being produced for cooking purposes with very simple equipment (Fig. 15.6). Biogas is produced from a wide range of organic materials of plant and animal origin (Table 15.9). Mixing animal and plant wastes in a biogas system has the advantage of complementing the highly nitrogenous animal waste with the nitrogen-poor, but carbon rich plant residue (Hashimoto, 1983). Conversion of biomass into methane involves a number of steps. The following three groups of bacteria are needed for the conversion of biomass into methane

1. Hydrolytic Bacteria

Bacteria included in this group include species of *Bacillus, Clostridium, Escherichia* and *Ruminococcus*. These bacteria degrade cellulose and proteins to form hydrogen, CO_2, propionate, butyrate, fatty acids and ethanol.

2. Acetogenic Bacteria

These bacteria convert most of the products produced by fermentive or hydrolytic bacteria into acetate. Some of the common acetogenic bacteria are: *Desulphovibrio* and *Syntrophomonas*.

3. Methanogenic Bacteria

Methanogenic bacteria act upon the products produced by the action of second group of bacteria resulting in the release of methane.Bacteria included in this

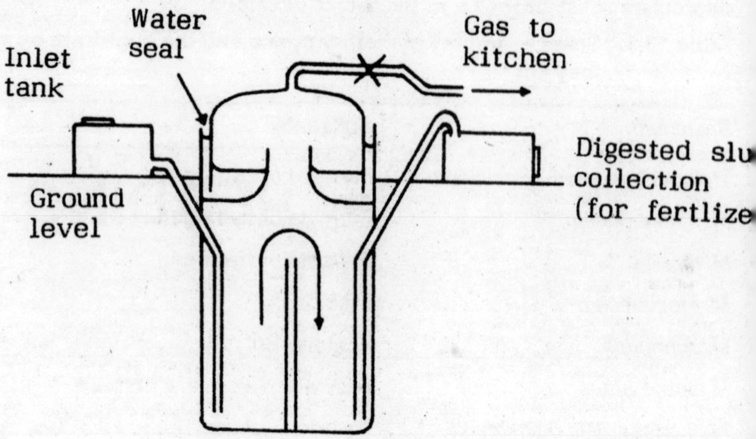

Fig. 15.6. Domestic scale biogas generator

group are listed in table 15.8.

Table 15.9. Some examples of organic materials used for methane production

Waste	*Examples*
Crop residues	Bagasse, weeds, straw
Aquatic plants	Water hyacinth, macrophytes, seaweeds
Animal	Cattle dung, sheep and goat droppings, fishery wastes, poultry droppings
Human	Faeces, urine
Agroindustrial	Oilcake, rice bran, wastes from fruit and vegetable processing

The gas yield from anaerobic digesters depends upon two main factors: (i) nature and concentration of substrates and (ii)temperature of incubation. The methane content ranges from 60-75% with CO_2 accounting for 25-30% and the remaining gases being hydrogen and nitrogen. In India and China biogas is produced from cowdung and other wastes whereas in countries like USA, water hyacinths are used. The type of anaerobic digesters which are presently being used for production of methane include batch digesters, stirred tank digesters, plug flow digesters, upflow anaerobic sludge blanket reactors, anaerobic filters and expanded or fluidized bed reactors.

FURTHER READING

Adams, R.L.P., J.T. Knowles, and D.P. Leader 1987. The Biochemistry of nucleic acids. Champman and Hall, London.

Ammirato, P. V. and D. J. Styer 1985. Strategies for large scale manipulation of somatic embryos in suspension culture. In: Biotechnology in Plant Science. Relevance to Agriculure in the eighties (M. Jaitlin, P. Day and A. Hollaender. Eds.), Academic Press, New York, pp 161-178.

Augustine, P.C., H.D. Danforsh and M.R. Bakst 1986. Biotechnology for solving agricultural problems. Martinus Nijhoff Pub., New York.

Bajaj, Y.P.S. 1986. Biotechnology in Agriculture and Forestry. Martinus Nijhoff Pub., The Netherlands.

Bayliss, M.W. 1980. Chromosomal variation in plant tissues in culture. Inter. Rev.Cyto. (Suppl.) 11A: 113-144.

Bhojwani, S.S. and M.K. Rajdan 1983. Plant Tissue Culture:Theory and Practice. Elsevier, Amsterdam.

Bonga, J.M. and D.J. Durzan 1987. Cell and Tissue Culture in Forestry. Martinus Nijhoff Pub., Dordrecht.

Brown, C.M., I. Campbell and F.G. Priest 1987. Introduction to Biotechnology. Blackwell Scientific Publications, Oxford.

Brown, T.A. 1986. Gene Cloning. Van Nostrand Reinhold, New York.

Bulock, J. and B. Kristiansen 1987. Basic Biotechnology. Academic Press, London.

Burke, D.T., G.F. Carle and M.V. Olson 1987. Cloning of large segments of exogenous DNA into yeast by means of artificial chromosome vectors. Science 236: 806-812.

Chopra, V. L. and Anwar Nasim 1990. Genetic Engineering and Biotechnology- Concepts, Methods and Applications. Oxford & IBH Publishing Co., New Delhi.

Cocking, E.C. and M.R. Davey 1987. Gene transfers in cereals. Science 236: 1259-62.

Crocomo, O.J., W.R. Sharp, D.A.Evans, J.E., Bravo, F.C.A. Tavares and E.F. Paddock 1986. Biotechnology of plants and microorganisms. Ohio State University, Columbus.

Crueger, W. and A. Crueger 1984. Biotechnology: A Textbook of Industrial Microbiology. Sinauer Assoc., Sundarland, Massachusetts.

Dale, J. W. 1987. Cloning in bacteriphage lambda. In Techniques in Molecular Biology, Vol.2. (J. M. Walker and W. Gaastra,Eds) Room Helm, London.

Dunstan, D.I. and T.A. Thorpe 1986. Regeneration in forest trees. In: Cell Culture and Somatic Cell Genetics in Plants. Vol. I(I.K. Vasil,Ed). Academic Press, New York pp. 223-242.

Evans, D.A. 1983. Agricultural applications of plant protoplast fusion. Biotechnology 1: 253- 261.

Evans, D.A. and W.R. Sharp 1986. Application of somaclonal variations and crop improvement. Nijhoff, Dordrecht.

Evans, D.A., W.R. Sharp, P.V. Ammirato and Y. Yamada 1983. Handbook of Plant cell culture (series of volumes). Collier Macmillan, London.

Fersht, A. 1985. Enzyme structure and mechanism. Freeman, San Francisco.

Foeler, M.W. 1986. Industrial applications of plant cell culture. In: Plant cell culture Technology,(Yeoman, M.M. Ed.) Blackwell, Oxford.pp 202-227.

Freifelder, D. 1987. Microbial Genetics. Jones and Bartlett, New York.

Grierson, D. and S. Covey 1986. Economic aspects of Biotechnology. Cambridge University Press, Cambridge.

Gelvin, S. B., R. A. Schilperoort and D.P.S. Verma 1989. Plant Molecular Biology. Kluwer Academic Pub., U. S. A.

George, E. F. and P.D. Sherrington 1984. Plant propagation by tissue culture. Exegetics Ltd, Eversley, England.

Gleba, Y.Y. and K.M. Sytnik 1984. Protoplast fusion.Springer-Verlag, Berlin.

Hackett, P.B., J.A. Fuchs and J.W. Messing 1984. An Introdution to Recombinant DNA technique. Basic experiments in gene manipulation. Benjannin/ Cummings Publishing Co., California.

Hacking, A. J. 1986. Economic aspects of Biotechnology. Cambridge University Press, Cambridge.

Hammerling, G.J., U. Hammerling and J.F. Kearney (eds) 1981. Monoclonal antibodies and T-cell hybridomas : Perspective and technical advance. Elsevier, Amsterdam.

Higgins, I.J., D.J.Best and J. Jones 1985. Biotechnology. Blackwell Scientific, Oxford.

Jain, H.K. 1988. Plant genetic resources and industry. Trends Biotechnol. 6: 73-77.

Kenneth, H.F., T.J.McKearn and K.B. Bechtol 1980. Monoclonal antibodies : A new dimension in biological analysis. Plenum Press, New York.

Lewin, B. 1990. Genes IV. Oxford Univ. Press, London.

Lindsey, K. and M.G.K. Jones 1990. Plant Biotechnology in Agriculture. Prentice Hall, New Jersey.

Maniatis, T., E. Fritsch and J. Sambrook 1982. Molecular cloning : a laboratory manual. Cold Spring Harbor Laboratory, New York.

Mantell, S.H. and H. Smith 1983. Plant Biotechnology. Cambridge Univ. Press, Cambridge.

Mantell, S.H., J.A. Matthews and R.A. Mckee 1985. Principles of Plant Biotechnology:An Introduction to Genetic Engineering in Plants. Blackwell Scientific Publications, Oxford.

Marx, J.L. 1989. A revolution in Biotechnology. Cambridge Univ. Press, London.

Moses, V. and R.V. Cape 1991. Biotechnology. The Science and the Business. Harward Academic Publishers, London.

Natesh, S., V.L. Chopra and S. Ramachandran 1987. Biotechnology in Agricuture. Oxford and IBH Publishing Co., New Delhi.

Neumann, K.H., W. Barz and E. Reinhard 1985. Primary and Secondary metabolism of plant cell culture. Springer - Verlag, Heidelberg.

Old, R.W. and S.B. Primrose 1985. Principles of gene manipulation : an Introduction to Genetic Engineering, 3rd ed., Blackwell Scientific Publications, Oxford.

Prave, P., U. Faust, W. Sitting and D.A.Sukatsch 1987. Basic Biotechnology. VCH Verlagsgesellschaft, FRG.

Raghavan, V. 1986. Embryoge sis in Angiosperms. A Developmental and Experimental study. Camoridge Univ. Press, Cambridge.

Rees, H., R. Riley, E.L. Bresse and C.N. Law 1981. manipulation of genetic systmes in plant breeding, Phil. Trans. R. Soc. Lond. B 292, pp. 399-609.

Reinert, J. and M.M. Yeoman 1982. Plant cell and Tissue culture : a laboratory Manual. Springer - Verlag, Berlin.

Rusell, G.E. 1988. Biotechnology of higher plants. Intercept Ltd, England.

Samal, J. 1986. Somaclonal variations and crop improvement. Martinus Nijholf, MA.

Sasson, A. 1988. Biotechnologies and development. Unesco/CTA. Paris.

Sharples, F.E. 1987. Regulation of products from Biotechnology. Science, **235**: 1329-1332.

Skoda, J. and H. Skodova 1987. An outline for food chemists and biotechnologists. Elsevier, Amsterdam.

Smith, J.E. 1988. Biotechnology. 2nd edition. Edward Arnold, London.

Vasil, I.K., W.R.Scowcroft and K.J. Frey 1982. Plant improvement and somatic cell genetics. Academic Press London.

Vasil, I. K. 1984. Cell culture and somatic cell genetics of plants. Academic Press, Orlando, Florida.

Vasil, I.K. and V. Vasil 1980. Clonal propagation. Internat. Rev. Cytol. **21**: 145-173.

Wareing, P.F. and I.D.J. Phillips 1981. Growth and Differentiation in Plants. 3rd ed., Pergamon Press, New York.

Winnacker, E.L. 1987. From genes to clones: An introduction to gene technology. VCH Publishers, New York.

Wiseman, A. 1987. Principles of Biotechnology. Chapman and Hall, New York.

Yeoman, M.M. 1986. Plant cell culture Technology. Botanical Monographs. Vol.23. Blackwell Scientific Publications, Oxford.

GLOSSARY

Acetobacter: A genus of gram-negative, aerobic bacteria of uncertain affiliation comprising ellipsoidal to rod-shaped cells as singles, pairs, or chains; they oxidise ethanol to acetic acid.

Airlift fermenter: Vessel in which bioconversion process takes place, sparged gas being the only source of agitation.

Amino acid: An organic acid that contains an amino ($-NH_2$) group; the building block (micromolecule) of protein.

Aneuploid : Cell or organism possessing a chromosome number other than the haploid number or an exact multiple of it.

Androgenesis: Development of plants from male gametophyte.

Anther culture: Culture of anthers (or pollen grains) on a suitable medium for production of callus or haploid plants.

Antibiotic: A chemical substance, produced by microorganisms and synthetically, that has the capacity in dilute solutions to inhibit the growth of, and even to destroy, bacteria and other microorganisms.

Antibody: Protein produced in the body in response to an antigen and capable of specifically binding with the antigen that stimulated its formation.

Antigen: Substance that stimulates an immune response when introduced into the body.

Attenuated vaccine: Whole pathogenic organism treated with chemical, radioactive, or other means to render it incapable of producing infection.

Autoradiograph: Photograph showing location of radioactivesubstances in cells or tissues, otained by exposing photographic emulsion in dark to radioactive emissions from the preparation, followed by development of

Auxins: A group of hormones of plants which are synthesised by growing tips of stems and roots and regulate many aspects of plant growth e.g. IAA.

Axillary: Developing in the axil of leaves.

Bacteriophages: Viruses that multiply in bacteria.

Base sequence: The specific order of purine and pyrimidine bases in a polynucleotide.

Batch culture: A suspension culture in which cells grow in a finite volume of nutrient medium and follow a sigmoid pattern.

Batch processing: Method in which bioreactor, for example, is loaded with raw materials and microorganisms, and the process is run to completion, at which time products are removed.

Bioassay: Method of measuring the concentration or activity of a substance by determining its effect on a living organism.

Bioconversion: Usage of biocatalysts for chemical processes.

Biodegradable: Organic compounds that can be broken down by micro-organisms.

Biofertilization: Nutrient enrichment of soil as a result of microbial activity.

Bioreactor: Apparatus used for bioprocessing.

Biosynthesis: The formation of organic compounds by living organisms.

Biomass: Weight of organisms per unit area. In biotechnology the term biomass refers to cellulose used for the production of chemicals to be used for energy.

Callus: A tissue consisting of dedifferentiated cells generally produced as a result of tissue wounding or of culturing tissues in the presence of auxins in particular.

Cancer: Uncontrolled proliferation of abnormal host cells.

Cell culture: Culture of single cells or groups of similar cells.

Cell differentiation: The process whereby descendants of a common parental cell achieve and maintain specialization of structure and function.

Clone: Individuals obtained from a single plant through asexual reproduction.

Cloning: Multiplication methods via clones.

Complementary DNA: DNA synthesised by reverse transcriptase from an RNA template.

Concatamer: DNA made up of repeated unit length DNAs.

Cosmid: A vector for carrying large DNA fragments into host cells.

Cultivar: A category of plants of the same species, but with a characteristic phenotype, generated through cultivation by man.

Culture medium: Nutrient material that supports cell growth.

Cybrid: Plant or cell which is a cytoplasmic hybrid produced by fusion of a protoplast and cytoplast.

Cytokinins: A group of hormones with stimulatory effect on divisions of plant cells e.g. kinetin and zeatin.

Cytoplast: Enucleated protoplast.

Denaturation: Alteration of material, usually protein, through heating or chemical treatment, so it loses its characteristic activity.

Differentiation: Modification in structure and function of the parts of an organism, owing to increase in specialization.

DNA probe: Sequence of DNA used to detect presence of particular nucleotide sequence.

DNA sequence: Order of nucleotide bases in DNA helix.

Electrophoresis: Separation of molecules based on their differing electrical charges.

Embryogenesis: A pathway of differentiation which is characterized by the formation of organised structures that resemble zygotic embryos.

Embryoids: Embryo-like structures produced as a consequence of differentiation processes such as embryogenesis and androgenesis.

Endonuclease: An enzyme that hydrolyses DNA by breaking bonds in the middle of the chain.

Enzyme: A protein which acts as catalyst to metabolic reactions.

Escherichia coli: A bacterium on which much genetic work has been done and which is a colon bacillus.

Eukaryotic: Refers to organism having membrane bound nuclei, genetic material organised into chromosomes and membrane bound cytoplasmic organelles.

Explant: A plant organ or piece of tissue used to initiate a tissue culture.

Fermentation: Decomposition occurring in organic compounds, usually carbohydrates, brought about by enzymes either directly or in microorganisms, resulting in the production of other compounds such as alcohol or lactic acid, often with heat and gases.

Gametoclonal variation: Variation observed among plants regenerated from gametic cultures.

Gene: Segment of a DNA molecule that contains information directing a cell to synthesize a specific protein product.

Gene library: Collection of DNA fragments (cDNA, genomic (gene bank) DNA) which have been cloned in a vector.

Genetic engineering: The experimental manipulation of DNA (or RNA) of different species producing recombinant DNA, which includes some genes from both species.

Genetic information: The information contained in a sequence of nucleotide bases in a DNA (or RNA) molecule.

Genome: Genome consists of all the chromosomes of a diploid species that are distinct from each other with respect to their gene content, and often morphology. Members of a genome do not pair.

Genotype: Genetic constitution of an organism.

Germplasm: Total genetic variability available to species.

Gram-negative: In the Gram stain, bacteria that do not retain the crystal violet dye after decolorization with alcohol.

Gram-positive: In the Gram stain, bacteria that retain the crystal violet dye after decolorization with alcohol.

Haploid: Having a set of unpaired chromosomes in each nucleus.

Heterokaryon: A cell in which two or more nuclei of unlike genetic makeup are present. Intron sequence of DNA interrupting the coding sequence of gene.

Homokaryon : Protoplast or cell with two, genetically identical nuclei, a product of somatic hybridization.

Hybrid: Offspring of two genetically-dissimilar parents.

Hybridoma: A method of creating pure and uniform antibodies against a specific antigen.

Immobilized enzyme or cell techniques: Techniques used for fixation of enzymes or cells on to solid support; immobilized cells and enzymes are used in continuous bioprocessing in bioreactors and upstream or downstream processing of materials.

Immunization: Administration of antigen to animal or organism to stimulate production of antibodies by that organism; also, administration of antigens, antibodies, or lymphocytes to organism to produce corresponding active, passive, or adoptive immunity.

Immunology: The science that deals with resistance to disease.

Interferon: Glycoprotein, important in immune function and thought to inhibit viral infections.

In vitro: (Latin, "in glass") Any process carried out in sterile cultures.

In vivo: (Latin, "in life") Any process occurring in whole organism.

Kilobase: A chain of nucleotides 1000 monomers in size.

Ligase: Enzyme used to join nucleic acid fragments together.

Meristem culture: Culture of apical meristems, particularly shoot apical meristem, for production of shoots and plantlets.

Messenger RNA (mRNA): RNA that serves as a template for protein synthesis.

Monoclonal antibodies: Those antibodies which are derived from a single source clone of cells and which recognise only one kind of antigen.

Morphogenesis: Developmental pathways in differentiation which result in the formation of recognised tissues.

Myeloma: Antibody-producing tumour cells.

Nick translation: A procedure for radiolabelling DNA *in vitro*.

Operon: A genetic unit consisting of adjacent genes that function coordinately under the joint control of an operator and a repressor.

Organogenesis: Type of morphogenesis which results in the formation of roots, shoots and /or floral organs.

Penicillin: The collective name for salts of a series of antibiotic organic acids produced by a number of *Penicillium* and *Aspergillus* species; active against most gram-positive bacteria and some gram-negative cocci.

Penicillium: A genus of fungi characterised by erect branching conidiophores having terminal tufts of club-shaped cells from which conidia are formed.

pH: A measure of the acidity of a substance, specifically, the logarithm of the reciprocal of the hydrogen ion concentration.

Plantlet: A small rooted shoot or germinated embryo.

Plasmid: Small, circular piece of extrachromosomal DNA in bacteria. Some plasmids can integrate into a bacterial chromosome.

Prokaryotic: Refers to organisms in which the nuclear material in not enclosed in a nuclear envelope and the genetic material is not organised into chromosomes and which lack membrane bound cytoplasmic organelles e.g. blue green algae, bacteria.

Protoclonal variation: Variation observed among plants regenerated from protoplast cultures.

Protoplast: A cell which lacks a wall.

32P: A radioactive isotope of phosphorus which emits strong beta particles and has a half life of 14.3 days.

Recombinant DNA (rDNA): Hybrid DNA produced by joining pieces of DNA from different organisms together *in vitro*.

Recombinant DNA technology: Use of recombinant DNA for specific purpose, such as formation of product and study of gene.

Restriction endonuclease: Enzyme that cleaves DNA at specific nucleotide sequences, often leaving single-stranded regions ("sticky ends").

Restriction enzyme: Enzyme that cuts DNA at specific DNA sequence.

Reverse transcriptase: Enzyme that polymerizes a molecule of DNA using RNA as the template for determining sequence.

Single cell protein (SCP): Cells, or protein extracts, of microorganisms grown in large quantities for use as human or animal protein supplements.

Somaclonal variation: Variation observed among plants regenerated from somatic cultures.

Somatic embryogenesis: Development of embryos from somatic cells in culture whose structure is to zygotic embryos found in seeds and with analogous embryonic organs such as cotyledons or cotyledonary leaves.

Somatic hybridization: A technique of fusing protoplasts from two contrasting genotypes for production of hybrids or cybrids which contain various mixtures of nuclear and/or cytoplasmic genomes, respectively.

Splicing: A stage in RNA processing in which introns are removed and exons are joined to form a continuous coding sequence of RNA.

Stirred tank bioreactor: Agitated vessel in which bioprocess takes place; mixing is provided by mechanical action of impeller/agitator.

Synthetic seeds: Embryoids preserved in a thin layer of calcium alginate.

Template: The macromolecular mold for the synthesis of another macromolecule.

Ti-plasmid: Plasmid from *Agrobacetrium tumefaciens*, used as plant vector.

Tissue culture: Culture of plant cells and tissues *in vitro* on artificial media.

Totipotency: Potentiality or property of a cell to produce a whole organism.

Transcription: A process involving base pairing, wherby the genetic information contained in DNA is used to order a complementary sequence of bases in an RNA chain.

Transduction: The transfer of bacterial genes from one bacterium to another by a bacteriophage particle.

Transformation: The genetic modification induced by the incorporation into a cell of DNA purified from cells or viruses.

Transgenic organisms: A genetically manipulated organism containing in its genome one or more inserted genes of another species.

Translation: The process where the genetic information present in an mRNA molecule directs the order of the specific amino acids during protein synthesis.

Vaccine: A preparation of a pathogenic microorganism which is treated to destroy its virulence when introduced into the body, causing the body to produce antibodies and so conferring immunity.

Vector: An agent of transmission.

Viruses: Infectious disease causing agents, smaller than bacteria, which always require intact host cells for replication and which contain either DNA or RNA as their genetic component.

Vitamin: An organic compound essential in small amounts for the normal metabolism of other nutrients.

SUBJECT INDEX